D0992169

KNOWING, DOING, AND SURVIVING

KNOWING, DOING, AND SURVIVING

COGNITION IN EVOLUTION

LIBUSE LUKAS MILLER

A WILEY-INTERSCIENCE PUBLICATION

JOHN WILEY AND SONS
New York • London • Sydney • Toronto

Library of Congress Cataloging in Publication Data:

Miller, Libuse Lukas.
Knowing, doing, and surviving.

"A Wiley-Interscience publication."
1. Knowledge, Theory of. 2. Man. 3. Analysis (Philosophy). 4. Civilization—Philosophy. I. Title.

BD161.M46 121 73-1198
ISBN 0-471-60512-3

Printed in the United States of America

10 9 8 7 6 5 4 3 2 1

To my high school history teacher

Leslie Lewis Bradley

who started me on the thinking habit when I was still too young to know where it would lead

"O the mind, mind has mountains; cliffs of fall
Frightful, sheer, no-man-fathomed. Hold them cheap
May who ne'er hung there."

Gerard Manley Hopkins

Preface

In the process of constructing civilizations, Humpty Dumpty has had a great fall, and the pieces into which he has broken have been increasingly pulled apart into two piles that are assumed to be opposed and incompatible, as some of the labels attached to them indicate: cognitive/emotive, objective/subjective, rational/nonrational, factual/evaluative. This book is an attempt to put Humpty Dumpty together again, by showing that the pairs of supposed opposites are in continuous interaction with, and interdependence on, each other, constituting a single indivisible process of knowledge, in the animals as well as in man.

The entire outlook of this book is evolutionary, which means that it tries to do justice both to our continuities with the other animals and to our undeniable differences from them. The book is as "empirical" as I was able to make it, but that word must be taken in a wider sense than is customary in scientific or technological circles. It is also unavoidably interdisciplinary, since I could not treat the many different kinds of knowledge, ranging from exact science to exact poetry, without entering many different fields. Although I have avoided neologisms, I have not been able to exclude entirely some of the specialized jargon of the different disciplines represented. The book is intended both for the specialized reader and for the general reader who is willing to do some exploratory work.

Naturally, I must beg the leniency of the former in his specialty, and the patience of the latter in covering so much ground.

I cannot begin to list the people to whom I am indebted, since college days, for opening my eyes to the cognitive possibilities and limitations in the widely separated fields brought together here. But I owe a special debt of gratitude to friends who have read all or parts of the manuscript for me, and made many helpful criticisms and suggestions: Kenneth E. Boulding, Robert W. Daniel, Richard F. Hettlinger, William E. McCulloh, Leslie Paul, Norman Pittenger, John F. Porter, John Crowe Ransom, and Charles Joseph Stoneburner. They are not to be held responsible for its deficiencies.

Thanks are also due to those members of the Mathematics and Physics Departments at Kenyon College who helped me formulate in ordinary language certain expressions in the discussions on order and chaos; and, most of all, to my husband, Franklin Miller, for his loving patience and technical assistance in acting as my literary secretary.

LIBUSE LUKAS MILLER

February 1973
Gambier, Ohio

Acknowledgments

The Jerome Rothstein quotation (Chapter 2, Ref. 5) is from *Physics Today,* September 1962.

The Joseph Church quotation (Chapter 3, Ref. 10) is from *Language and the Discovery Of Reality* by Joseph Church. Copyright © 1961 by Joseph Church. Reprinted by permission of Random House, Inc.

The Wallace Stevens quotation (Chapter 3, Ref. 13) is Copyright 1942 by Wallace Stevens. Reprinted from *The Collected Poems of Wallace Stevens,* by permission of Alfred A. Knopf, Inc., and by permission of Faber and Faber, Ltd., from the *Collected Poems.*

The Karen Horney quotations (Chapter 3, Refs. 17–19) are reprinted from *Neurosis and Human Growth* by Karen Horney, M.D. By permission of W. W. Norton & Company, Inc. Copyright 1950 by W. W. Norton & Company, Inc. Also with permission of Routledge & Kegan Paul, Ltd., London.

The quotation from Plato's *Symposium* (Chapter 7, Ref. 5) is from *Great Dialogues of Plato,* as translated by W. H. D. Rouse. Copyright © 1956 by John Clive Graves Rouse. Reprinted by Arrangement with The New American Library, Inc., New York, New York.

The Donald A. Stauffer quotation (Chapter 9, Ref. 1) is reprinted from *The Nature of Poetry* by Donald A. Stauffer. By permission of W. W. Norton & Company, Inc. Copyright 1946 by W. W. Norton & Company, Inc.

Additional acknowledgments are referenced at the points in the book at which quotations occur.

Contents

KNOWING, DOING, AND SURVIVING

PART ONE

The Theory

One Toward a Unifying Theory of Knowledge

I

How does one start a theory of knowledge? To question the contents of our own consciousness, in order to discriminate the parts that qualify as knowledge from the parts that do not, seems to require that we be able to stand somewhere outside our own consciousness, that we can survey these contents like a panorama laid out before us, and that we be equipped with a syllogism which runs: All knowledge has the characteristics XYZ. ABC claims to be knowledge: Therefore ABC must have the characteristics XYZ, or it does not qualify. After attempting that, we feel we have not only destroyed the basic unity and certainty of our consciousness but have also started an endless argument about XYZ, ABC, and the meaning and use of the word "knowledge" in our language.

We cannot start a theory of knowledge from scratch; and there is our first problem. In beginning to construct a theory of anything else—for example, a theory of the formation of galaxies—we are allowed to assume that there are possibilities of knowing and communicating believed in and trusted by all parties to the discussion. We must be on guard that these conditions are observed, so that the would-be theorist need only describe the facts and theories available up to that point and show how the new theory explains them better than previous theories. But a theory of knowledge is concerned with just those possibilities and conditions of cognition and communication which all other theories take for granted in order to get started—therefore, how can it get started at all? And how can it avoid traveling in circles—always having to assume at least part of the very thing that is to be proved or demonstrated? For surely if we write down an English sentence about knowledge, we have to assume that this little item of would-be knowledge about knowledge, the content of the proposition, is at least possible and is communicable by means of an English sentence. And we even have to assume that this item of knowledge would not suffer seriously if an entirely different set of words or sounds were substituted for what we have written, that is to say, if the sentence were translated into any number of different languages. Thus we can hardly present our theory of knowledge as if we were superhuman, external observers, speaking from some point outside the confines of language, for we give away our position right from the start, by not being able to talk about knowledge without assuming some kind of knowledge.

Ideally, a theory of knowledge should precede all other theories, because only a theory of knowledge could validate the grounds on which all other theories rest. Or, alternately, it could invalidate them—and thereby it could cut the ground from under them. Didn't positivism try to cut the ground from under metaphysics? But chronologically and historically, things do not happen ideally, and we know that a serious concern with theory of knowledge did not arise in the history of philosophy until the Renaissance, when new discoveries in astronomy and geography *forced* a radical questioning of the existing body of supposed knowledge. And then a curious thing happened, which has been repeated in roughly similar form at least twice since the sixteenth century. The new findings were not used simply to correct the previous ideas in the field in which they applied—in this case, astronomy, the heliocentric view of the solar system replacing the geocentric. No, the new findings were felt somehow to upset the whole view of reality and to bring up for questioning the very nature of knowledge itself. It was as if the thinkers were saying to themselves: there must be something wrong with our idea of knowledge and with our notions of where and how it can be obtained. Otherwise, how could we have been so mistaken for so long?

About four hundred years later, at the beginning of the twentieth century, a similar epistemological flurry followed the need to replace the Newtonian mechanics by the quantum and relativity pictures of physical reality. And previously, although much more slowly, there had been the need to replace the theory of spontaneous generation of biological forms with the theory of evolution. In both cases, what seemed to be purely local disturbances in very specialized fields of knowledge spread out in all directions to become tremors of doubt that threatened to shake the foundations on which human enterprise and interest had rested more or less securely for thousands of years. How these scientific discoveries affected the field of religious beliefs we need not rehearse here once again. We are now still in the epistemological backwash following the theory of evolution.

Theory of knowledge is a dangerous subject, just because it is so closely connected with, and interpenetrated by, theories about man and theories about God. And it is even possible that the theory of knowledge is a subject that never should have been raised, since it certainly destroys the innocent, uncritical, naïve realism of the learning child in his world of known and knowable things, always open to new discovery and correction. Isn't it conceivable that given enough time, and patient learning, and absence of preconceived notions, all these great mistakes or errors, such as the geocentric system, would have quietly corrected themselves, in due course, without that terrible questioning of all things that resulted from putting knowledge itself on the rack? But alas, the adult mind must have its finished picture! So it prematurely finishes the picture, and thereby almost guarantees another upheaval at the next great discovery, and another torturing of the subject of knowledge, and another self-torturing of man about his capacity as knower. But why must the adult mind have its finished picture? That is one of the questions we must explore in this essay.

Let us not approach our subject with that most pervasive and subtile of self-delusions, namely, that we ourselves are without preconceived notions or basic cultural assumptions. None of us was raised in a cultural or historical vacuum, and the very language we use is philosophically "loaded" in its structure and its category-texture. But we do not have time or space here for a general examination of basic cultural assumptions. The brief summary of my own in Appendix Two allows me to refer the reader to them at those points in the argument where I am conscious of their influence in my presentation. The reader can ignore them; or, if he reads them, he will surely partly or even utterly disagree with them. But at least he will become more conscious of his own basic cultural convictions by not being able to agree with mine.

II

We know, then, that our procedure in starting to construct a theory of knowledge can be neither strictly inductive, nor strictly deductive, and that it must rest on presuppositions that constantly threaten to make it circular. Not strictly inductive, because the single items of supposed knowledge from which we might inductively abstract a generalization have not yet been shown to constitute knowledge; and not strictly deductive, because the relation between logical implication and reality has not yet been established. There is nothing for it except to take a deep breath and plunge into a going concern that constantly assumes itself as it goes along. But we can help ourselves at the outset by asking a few anticipatory questions about what we are looking for, as a diver might try to picture in his mind the object of his search before he lets himself down to the dark sea floor.

We are looking for a theory of knowledge that will put Humpty Dumpty together again, and so we must first of all try to discover what the pieces are and where they are to be found. Unlike the deep-sea diver looking for the pieces of an ancient vessel or a lost city, we do not have to worry about the pieces having been widely scattered by storms and seismic upheavals. Our pieces are all in the same place, in the consciousness of every living man and woman; and, moreover, they are obligingly reborn for us with something of their pristine unity with every new generation of children. The falling apart is a cultural and historical phenomenon, not a biological one.

In fact, let us admit that the falling apart is a Western, even a European phenomenon, due to the avalanche of knowledge that has forced division of labor and specialization on all educated people. That ideal of the Renaissance, the well-rounded man who had a bird's-eye view of all existing knowledge and yet was active in the world (Leonardo da Vinci), is an impossibility today. We are all so proud that we have so much education; but it is a consciousness-shredding machine, proliferating specialized languages to such a degree that even within specialized disciplines, only the sub-sub-sub-specialists can talk to one another.

Today the words "interdisciplinary approach" express the need and hope for the achievement of some degree of reunification and order out of the chaos of fragmentation brought on by specialization. With the best intentions in the world, however, this approach is more easily mentioned than applied. For the pieces into which Humpty Dumpty has fallen are not just lying there passively, waiting to be reassembled. Some of the pieces are actively fighting each other, or are "not on speaking terms," except for trading insults. In philosophy, for example, people who read Wittgenstein will not touch Heidegger, and vice versa. In psychology, existentialists will not touch "animal" psychologists; Heidegger has even said that "scientists do not think"! All parties have the fuzziest notions about art, except art-cognitivists, and they regard "psychologizing" as beneath contempt. But even in those sciences or social studies that are partially quantifiable, or subject to statistical analysis, there are schools and schools of interpretation. We have only to think of the manner in which Americans tend to deal with every problem by setting up a commission of experts, and then ignoring the experts' recommendations because the findings were neither conclusive nor unequivocal. They never ask, Could they be? The situation is analogous to that of the United Nations in trying to produce international cooperation: what faction, which party or junta or dictatorship, shall be listened to as truly representing each discipline?

It begins to look as if the pieces of Humpty Dumpty in the opposed pairs: cognitive/emotive, objective/subjective, rational/nonrational, factual/evaluative, and so on, had each decided to go into business for itself, producing a new concept of modern man in which itself is the most important unifying factor. The excuse for this is that modern man is so different from ancient man (or Medieval, or Renaissance, or Enlightenment man) that he can no longer use those religious or political or tribal concepts that served them for unification.

In this essay we are looking for a theory of knowledge that will not leave out any of these pieces and that will assign to each one its proper functioning in relation to the whole, thereby demonstrating that modern man is not so different from ancient man after all. Certainly the external world is different, and modern man thinks he is different. Right now, in some quarters, he thinks he is some kind of biological computer. But paleontologists tell us that this "biological computer" has not changed since Cro-Magnon times, and they have some existing Stone Age people to prove it. A brain surgeon could not tell the difference between one of these and a self-styled modern man. So then the differences must be psychological and cultural. In using computer-talk, let us never forget GIGO: garbage in, garbage out.

There is an element of urgency in our quest. Paleontologists and evolutionists recognize that man is the problem-solving animal. It was by being the problem-solving animal that he outwitted and outsurvived the other animals in the long evolutionary struggle. But now this problem-solving animal, armed with the most destructive combative and exploitative techniques in the history of evolution, is preparing to terminate his own species and to poison the planet for all other animal life, because he cannot solve the problem of his own character. In fact, he has even figured out some solutions to the problems he has created by his excess of power and knowledge in the field of technology—the solutions are already "on the drawing boards." But what good will that do if his failure to solve the problem of human character means that not enough people will want to put the solutions into action soon enough—before it is too late, if it is not already too late?

We have only to set these problems down in capsule form—pollution, overpopulation, war, and tribalism—to see that they are all interrelated and that they are all ancient, not new.* They have been with man since at least Neolithic times, the period from which we count him as definitely human (Cro-Magnon), as against prehuman. But they were then mild, attenuated, and always solvable by new inventions or by the simple solution of moving to another place. It is only man's technological "conquest" of nature with science that has sky rocketed these problems into explosive proportions, so that

* Of these four, tribalism is the most ambiguous, since in both its ancient and its modern forms, such as nationalism, it brings out the best in man as well as the worst.

they now hang over the whole species and the planet like four apocalyptic time-bombs. And we have only to consider the many solutions proposed for these problems to realize that technology alone will not solve any of them. Only a change in the character of man will solve them, and this must be a cultural change, since *there simply will not be the evolution-time needed for a biological change.* Fortunately, man has shown himself to be culturally quite flexible, as the great variety of existing cultures testifies. Furthermore, in the matter of changing man's character culturally, all the higher religions of mankind are agreed on what must be done—that is, all those religions with ethical teachings on man's proper conduct toward his fellow man. Whatever their theological or philosophical differences, all are agreed that man needs to be delivered out of *some* aspects of his primeval Stone Age character and to cultivate and encourage *other* aspects and possibilities in that character, in order to achieve true humanity. All have something to contribute, if they will only set their sights on this goal.

Two The Primacy of Vision in Human Ideas of Order and Chaos

I

In our first step in the construction of a theory of knowledge, let us examine the sensory situation in man, with special reference to the title of this chapter, the primacy of vision in the animal man, and its effect on his ideas of order and chaos. Again we must plunge into a going concern, one of enormous complexity in its actual components at the present time, but even more complicated by thousands of years of speculation on the meaning and function of sensory experience, on both the "common sense" and the philosophical levels. Idealistic philosophers are always telling us that there is something "low" about the senses, not only because we share them with the other animals, but more because the senses can't be trusted: they change, they sometimes delude us, they may try to enslave us! On the other hand, doctors and lawyers think so highly of the senses that they make them the virtual touchstones of sanity: anyone who hears what other people don't hear or sees what other people don't see, is under suspicion!

Furthermore, both the doctors' evidence (for diagnosing a disease) and the lawyers' evidence (for circumstantial ability to commit a crime) are largely sensory. Witnesses in a court of law have to swear that they saw, heard, or touched the accused, at the proper time and place, not just that they somehow knew or believed he was there.

The doctors and lawyers are on the side of common sense. But when modern philosophers who claim to be on the side of science want to talk about the senses, instead of telling us what the sciences say about the different senses that have been investigated, they claim that we must begin with something they call "sense-data." These philosophers are usually linguistic analysts who are worried about the trustworthiness of types of sentences, and far be it from me to deny that a great deal can be learned from them about the difficulty of saying something that is absolutely certain, or merely probable, or potentially verifiable, in the present, past, or future. We take up the problems of language and philosophical analysis in Chapters Five and Eight respectively. Here I am simply noting that, just by starting on such a high level of abstraction as "sense-data," certain philosophers are creating certain problems that are built into abstractions per se; and quite unintentionally, they are fostering various delusions about man's sensory situation in the world.

For theory of knowledge, "sense-data" are supposed to be the most primitive ingredients, the simplest building blocks, out of which the more complex items of knowledge are built by some yet-to-be-investigated "logical" process inside the rational mind. The experimental evidence for this claim rests on a well-known armchair experiment that anyone can repeat by emulating what the philosopher does. The philosopher sits perfectly still in his armchair, thereby scientifically reducing the number of variables. Staring ahead of himself, he begins to describe what he sees in as "primitive"—that is to say, in as unidentified, uninterpreted, unselected and unpreferred—a manner as he possibly can. He lists shapes and colors, breaking them down into tiny parts, such as spots of color and bits of line. What's wrong with that? Nothing. But two wrong *interpretations* of this experiment are possible and are not always carefully guarded against. The first is the delusion that this experiment resembles what a scientist does in collecting "sense-data," since the scientist deliberately ignores everything except a selected specific item that the structure of the experiment was designed, at great labor and expense, to show. The second is the delusion that this experiment, which breaks down the visual field into unintelligible bits, resembles the process by which we go from our earliest childhood sensations to our earliest sensory perception of material objects. By running the process backwards, it may seem to demonstrate how children proceed from a pure sense-data-chaos to the knowledge of "things."

Now this armchair experiment does show something. It reveals that a grownup, by deliberately withdrwing his attention from the already-known visual objects before his eyes, can induce in himself a state of trance, fascination, curiosity, or reverie, in which he is *paying attention* only to spots of color and bits of line—but it is highly doubtful that *that* is the state (reversed) in which children get from their earliest sensations to their earliest sensory perception of material objects. When a child old enough to focus his eyes is handed his first apple, he does not see a fog of colored spots and bits of line where you and I see the apple—he probably sees the apple just as you and I, perhaps better, because that is how a focused eye functions. He does not yet have fused and integrated with this visual Gestalt certain other sensory patterns, but these he gradually acquires by *acting* upon the apple in various ways: its weight, its smoothness, its hardness, its good smell, its crunchiness, its good taste. And all this probably long before he hears the word "apple."*

* It may very well be that one reason theory of knowledge is so easily bogged down in verbal muddles is that it operates with such a vague and imprecise psychology. "Do we perceive material objects *directly,* or only

In this essay we seek help from psychologies on many levels—including animal, behavioral, Gestalt, learning theory, child development, and psychoanalysis. But that does not mean that we are going to escape from the problems occasioned by the use of abstractions. One of the functions of abstractions is to act as a kind of verbal shorthand for covering a lot of ground with a single word, and since we are going to cover a lot of ground, the dangers will be many. The alternative is using words as exact, as specific, and as concrete as possible. For instance, in describing sense perception in grownups we should say: he sees the spot and perceives that it is a star; he hears the sound and perceives that it is a fire siren; he touches the surface and perceives that it is icy; he tastes the flavor and perceives that it is apple; he feels the pain and perceives that it is toothache. In each case, a whole biography of experience lies between the sensation and the perception. And far enough in *that* direction lies the dissolution of all science.

But that does not excuse us from trying to pinpoint the delusive propensities of our two words, "sense" and "data," in relation to the sensory *situation* of man in the world, since we are about to compare it with the sensory situations of the other animals. First, sensory illusions. The fact that the several senses operate in us in such qualitatively different ways is a tremendous *advantage* that we deprive ourselves of gratuitously, the moment we abstract from the differences in favor of the generalized term "sensory." All the ancient and modern skeptical arguments about how we cannot tell the difference between dream and reality, or between illusion and reality, fall to the ground as soon as we take the trouble to observe how the ordinary man, the nonphilosophical, "common-sense"

---

sense-data?" For an amusing analysis of this question, and all the mischief it has caused, see J. L. Austin's *Sense and Sensibilia,* pp. 15–19. Professor Austin calls the word "directly" in this question "a great favorite among philosophers, but actually one of the less conspicuous snakes in the linguistic grass," and proceeds to demolish whatever possible meanings it could have.

man, actually goes about in daily life using his several sense organs to distinguish natural or artificial illusions from reality, and dreams from reality.

If I cannot convince myself by seeing alone that the scene before my eyes is real rather than illusory, more seeing will not help me—I need the independent corroboration of some other senses. Thus I prove the illusoriness of the cinema picture by touching the screen, I dispel the mirage by walking into it, I distinguish the artificial fruit from the real apple by smelling it and biting into it, I satisfy myself that the "heard" burglar is nonexistent by failing to see him or any evidences of his passage, and so on. I distinguish the dreaming state from wakefulness by something generally absent from the former—for example, my awareness of my weight against the bed, my position in the bed, the touch of the sheets and blanket against my body, the temperature of the air in the room. In fact, there exists an in-between state where, while being aware of all these evidences of wakefulness, I deliberately spin out or finish off some dream, but I do not call it sleep precisely because I know where I am and what I am doing. Suppose that now our abstracting skeptic protests by saying, "Ah, but it is the entire sensory bag of tricks that you call consciousness that cannot be proved not to be an illusion or a dream." We are obliged to answer that, according to this usage of the term, it must be a very peculiar illusion indeed that cannot be contrasted with any reality and that nevertheless permits within itself that the illusion–reality contrast arise, a sort of second-order illusion being distinguished from a second-order reality.

You hardly ever hear a physicist talking about the bent stick in water or the mirror-image as "illusions." To him, these phenomena are only vision operating under the optical conditions of refraction or reflection; and indeed it was their *curiosity* about these phenomena that led to far-reaching discoveries in optics, making order out of many seemingly chaotic "facts." But even the ordinary man, in the ordinary everyday business of getting around in the world and staying alive, is constantly "checking" the facts by using one sense perception to corroborate, to anticipate, to disregard, to hope for, or to guard against, a qualitatively different sense perception.

The pervasive delusion about man's sensory situation in the world that is promoted by the second word, "data," is the *passiveness of the receiver.* To start with the absurd first, our philosopher in his armchair will not be "given" anything, if it is nighttime and he has forgotten to pay his electricity bill—for he has to provide and pay for the light that will bounce off the objects and interact with his nervous system to produce the desired sensations. Ordinarily, the owner of the sensory equipment is quite well aware that he has to use all kinds of strategies to manage the content of his consciousness, so that mostly he will get the sensations he likes and desires, and mostly he will avoid the sensations he fears and dislikes. Whether he uses internal or external devices for seeking out or warding off (ignoring, paying attention, turning on or off sources of light or sound, avoiding, grasping, etc.), he knows he has to *do something;* that even his individual sense organs have to *work for* him, not against him; and that he has to take care of them, nurture them, give them rest, if his consciousness is to be bearable. In the morning (if he gets up healthy and refreshed) seeing, hearing, touching, tasting, smelling and kinesthetic sensations all seem to be tuned up and ready for action, like idling motors all set to seek adventure. In the evening when he is tired, the very same sensations that were sought out in the morning now strike him as an unwanted assault, and one by one he tries to ward them off. He seeks darkness, silence, comfort, inaction, sleep. To see that man is the actor here, not a passive recipient, we have only to watch the rituals that an insomniac performs to earn for his weary sensorium a few precious hours of unconsciousness.

"Nothing is in the mind that was not first in the senses," it used to be said in the olden times of the *"tabula rasa"* sensationalism—yes, nothing except several millions of years of naturally selected, inherited action-patterns, which help that mind to survive by means of, or in spite of, the sensations. If man really were a *passive recipient* in the process of sensation, he would have absolutely no incentive to get from sensation to sensory perception, for he could not change what was happening to him; and it is just this attempt to *change his consciousness* that forces him to deal with the world and to acquire sensory perceptions to help him. Neither is he an actor in a vacuum, but more like a stage director trying to make the play go his way. The built-in hedonism of small children, their living by the principle of seeking pleasure and avoiding unpleasure, is the best guarantee of their rapid and thorough education in perceiving the things of this world. Although this principle turns out to be inadequate for adult life, at the beginning it is indispensable. That is why material objects cannot be *logical constructs* made out of sense-data. They must rather be something like persistent, fused Gestalts of varied sensations, learned according to the child's earliest interactions with the child-world. And doesn't some of that primitive meaning of material object as anything-a-child-can-get-its-hands-around linger on all through life, even in sophisticated grownup talk about the atom as a material object? Or the galaxy as a material object? Or God as an object whose existence or nonexistence could be proved? Linguistically, we are given no choice: we must all begin where our first language began, in childhood; and a child cannot learn a language that does not deal with the child-world. If a grownup asks, "Do material objects exist, or only sense-data?" we should have to answer, "Well, if this fused Gestalt of sensations that I call a material object does *not* exist, we shall either have to find another meaning for 'exist' or we shall have to find a name for it other than 'material object.' " For "material object" is the *class name* I have learned to give to all those instances in which I have a persistent fused Gestalt of sensations from my interactions with the world. By such instances I mean "existing things" and certainly not "nonexisting things."

II

It has long been remarked by both psychologists and epistemologists that vision and hearing play the largest roles in our ideas of what knowledge is, and that vision predominates over hearing in that respect. Another way of putting it is to depreciate the role of touch, taste, and smell in the idea of knowledge, although admitting that smell, for example, plays a much larger role in the practical "knowledge of terrain" in some animal species than in man. Then, however, the subject is dropped, and we immediately lose the advantage of this insight by talking about sensory phenomena as if they were all alike.

Let us compare vision and hearing. We have only to reflect on certain everyday happenings in order to realize how overwhelming is the dominance of sight over hearing in the matter of delivering information to the human nervous system. The very pains that have to be taken by persons born blind, to learn compensatory reactions, indicate to the sighted people what a dangerous and impoverishing experience visual deprivation must be. But to get an idea of the *magnitude* of the difference between seeing and hearing, we must consider the reason why radio was invented before television. It was because the number of "bits" of information that had to be sent out on the carrier waves to reconstruct a recognizable picture at the receiving end was so much greater than the number of "bits" needed to reconstruct a recognizable sound, something like *hundreds of thousands* of times greater. Although the eyes and the ears are about equally acute when it comes to discriminating the just barely discriminable differences of stimulation, the objective fact seems to be that the world contains many more discriminable sources that send light waves to the eye, than discriminable sources that deliver sound waves to the ear. Vision, seeing, is a process that is kept much busier by the very nature of the world than is hearing, and life-imitating machines must reproduce this difference as best they can.

Again, consider how easy it is to achieve the restfulness of silence without plugging the ears, as compared with achieving the restfulness of relative blankness for the eyes, without closing the eyes. One has only to soundproof his room for the first, which is relatively easy; but for the second, short of denuding the house of discriminable objects, one must move to some place like the seashore, where the landscape is relatively simplified into large masses of sky, sea, and shoreline, a combination noted for its restful effect.

Or again, let us condider the sensory situation as it is revealed by the famous saying that a picture is worth ten thousand words. How exactly is a picture *worth* 10,000 words? Does this expression not mean that by looking at the picture you are saved the trouble of thinking up the 10,000 words that it would take to describe the content of the picture? The eye perceives at a glance innumerable complex relationships of form and color, depth and localization that could not be conveyed to a nonwitness without pages of laborious prose. Of course if the purpose is merely to identify a person or place, as in the investigation of a crime, all items except the relevant ones are ignored. But if any person just sitting in a room with a view from the window were assigned the task of describing exhaustively *everything* that he saw, every discriminable item and relationship, he would soon feel the oppressiveness of the task—for is there any limit to the new combinations and overlappings of design and relationship that he can bring himself to perceive if he will only pay attention to them? The task seems hopeless, because visual perception appears to be inexhaustible. However, since no one has tried it, we will not dogmatize about whether it can be done. The same richness of particularity lies behind what might be called the obverse of Confucius' saying, a feeling of the inadequacy or superficiality or inaccuracy of words, as when someone exclaims, "Words fail me when I try to tell what I saw!"

Now this oversupply of visual information fed by the sense of sight into the human nervous system turns out not to be quite the advantage that one would think it should be. Combined with the information from the other senses, this barrage of particulars shows man to be a creature embarrassed by sensory riches, a situation described by the authors of *A Study of Thinking* as apparently paradoxical:

> We begin with what seems a paradox. The world of experience of any normal man is composed of a tremendous array of discriminably different objects, events, people, impressions. There are estimated to be more than 7 million discriminable colors alone, and in the course of a week or two we come in contact with a fair proportion of them. No two people we see have an identical appearance and even objects that we judge to be the same over a period of time change appearance from moment to moment with alterations in light or in the position of the viewer. All of these differences we are capable of seeing, for human beings have an exquisite capacity for making distinctions.
>
> But were we to utilize fully our capacity for registering the differences in things and to respond to each event encountered as unique, we would soon be overwhelmed by the complexity of our environment. Consider only the linguistic task of acquiring a vocabulary fully adequate to cope with the world of color differences! The resolution of this seeming paradox—the existence of discrimination capacities which, if fully used, would make us slaves to the particular—is achieved by man's capacity to categorize. To categorize is to render discriminably different things equivalent, to group the objects and events and people around us into classes, and to respond to them in terms of their class membership rather than their uniqueness. Our refined discriminative activity is reserved only for

those segments of the environment with which we are specially concerned. For the rest, we respond by rather crude forms of categorial placement. (1)

Now it should be noted that only for man in his role as actor and responder is the seeming paradox resolved. The categories simplify the overwhelming complexity of his environment and thereby allow him to respond to, that is, to deal with, discriminably different things as if they were equivalent; but it is precisely for action that they are equivalent. For the contemplative attitude, that form of curiosity that simply looks at things as they are to see what they are, not to manipulate them, the discriminably different things are not equivalent—they remain what they are in their differences, the paradox is not resolved, and the overwhelming complexity remains as an almost palpable presence, charged with a variety of emotional content.

And as if it were not enough that the sense of sight is largely responsible for the density of particulars that surrounds man on all sides when he chooses to pay attention (and stands as a permanent background to all his activities when he chooses to ignore it), the imagination of man, when he assumes the attitude of contemplative curiosity, extends this multiplicity far beyond what is immediately perceivable to him—in fact, for the modern man, who no longer inhabits a cozy universe, to the imaginable extensions of space and time whose limits are precisely the unimaginable. The vertiginous feeling of chaos, the annihilating feeling of insignificance, the contrary feeling of exaltation before such an astounding spectacle, the feeling of awe, and even of terror, at being at the mercy of something so incomprehensible—these and many variations or combinations of emotions are the reward of the man who permits himself to adopt the "impractical" attitude of imaginative contemplation. Here also are to be found the feeling of "wonder" that for a certain type of mind is the beginning of all philosophy, and the feeling of the *mysterium tremendum et fascinans* that for a different type of mind is the beginning of all true religion. I say different types of mind, for as we shall see later, these two types of mind differ in what they think can be *done* about the immensity and complexity of the universe as it is presented by the power of vision extended in imagination. There are no doubt also types of mind without much imagination, and still others that have it but refuse to use it—for instance, people who see no practical results in it, and people who shun it precisely because of the feelings it arouses. The mere study of astronomy gives some people vertigo. Even the psalmist of old, in his much smaller universe, trembled at the disproportion: "When I look at thy heavens, the work of thy fingers, the moon and the stars which thou hast established, what is man that thou art mindful of him, and the son of man that thou dost care for him?"

And as if *that* were not enough, the imagination of man also turns in the direction of the infinitesimally small, to the fine structure of all things, to all that is invisible to the naked eye but presented to the eye of the mind by visual imagination, down to the molecules and the atoms and the subatomic particles—another overwhelming complexity to be added to that on the everyday level and to that on the astronomical level. Pascal, it will be remembered, characterized man as being placed between two immensities—the infinitely large, and the infinitesimally small—and he found this condition to be the occasion of great wonder.

With these two characteristics, then, visual imagination and contemplative curiosity, grownup man lives in a world different from that of animals and children. But let us not hasten to congratulate him on this possession, as the Greeks were wont to congratulate man on his possession of what they called "reason." For instead of making him happy, this simultaneous awareness of immensity and complexity creates in man the pressure, the need, the yearning, for some sort of unifying idea, principle, metaphor, symbol, or myth—anything that promises to make order out of the chaos of this immense totality! The discriminable differences simply will not be denied. The eye, that is, cannot deny what it itself discerns or perceives; and this burden of particulars, this overwhelming complexity, is passed on to the mind with a mandate for the mind to *find* the order in it. And the mind naturally turns toward knowledge, expecting to find through knowledge the order in or behind the chaos, for that is one of the defining characteristics of what man thinks knowledge to be: knowledge is that which makes order out of chaos.

When now we turn from this realm of dizzying infinities to the everyday world revealed in visual perception, we find a different story. It certainly seems to us that, at least on the size scale that man occupies, some kind of order is already built into the world, and it is again the sense of sight that shows it to us. However, this is true only if we do not pay attention to the tiniest particulars, the least discriminable differences; we must ignore them in favor of the more prominent and persistent fused Gestalts of sensations that we have learned to call "things" or "material objects." Here we are in a world in which every *thing* seems to have been made according to a *kind*, and it is the *kinds* that are more obviously different from one another than the members of each kind. Almost every object we see has a definite shape, and this shape either resembles or does not resemble other shapes, a fact that enables us to use our "capacity for categorizing" on it. But first it is the very *definiteness* of the shapes of things that suggests the idea of orderliness to man. There is nothing suggestive of chaos about the shape of a flower, a tree, a horse, a snow crystal—and the arrangement of the feathers on birds (where we might expect a tiny bit of chaos to creep in) amazes us by its fixed pattern, its unruffled orderliness, the same for all of its kind. Order suddenly seems to be everywhere in the natural world, in the definite shapes of things, and in the resemblances and nonresemblances that permit us to arrange them according to distinguishable kinds.

Fixed patterns and distinguishable kinds—these are the germinations of the ideas of order in man, and they are given to him predominantly by the sense of sight, just as much as his ideas of chaos. Small wonder that in the history of human consciousness the symbolic and metaphorical meanings that have accrued to the words "vision" and "light" are all in the areas of knowledge, power of mind, wisdom, truth:

> Of all archetypal symbols there is none more widespread and more immediately understandable than light, as symbolizing certain mental and spiritual qualities. Even in our current everyday vocabulary pertaining to mental phenomena there are many words and phrases that are products of earlier light metaphors: *eludicate, illuminate, clarify, illustrate, bright,* etc. . . .
>
> The image of light is thus extraordinarily well fitted to stand as the principal imagistic symbol for mind; *light* is the semantic vehicle while mind is the tenor. The organic relationship between the two is expressed by an ancient Zoroastrian saying, preserved by Porphyrius: "The body of Ahura Mazda is light, his spirit is mind." The nature of mind is elusive and ambiguous and no method of analysis can ever be adequate to a full understanding of it. But one thing we know about it indispensably—its power of discrimination. Whether in the field of action or in quiet contemplation the power to discriminate is the essential mark of mind, and this power above all is what *light* symbolizes. (2)

The primacy of vision in man's sensory situation thus gives him no choice but to be always in a state of tension, between the threat of chaos and the hope of order. The other senses, especially hearing, do this also to some extent, but on a scale of magnitude not nearly so great, especially in the extending medium of the visual imagination. The senses of taste and smell can hardly be said to give any fixed patterns, only distinguishable kinds, and not very many at that. However, if at this point we also ask, either impatiently or in contemplative curiosity, why does man have these emotional involvements with the ideas of order and chaos, why does chaos strike him as frightening and threatening, whereas order appears to him to be hopeful and reassuring, we get an unexpected answer from an unexpected quarter—from the probabilistic theory of entropy, as this is understood in the science of cybernetics. In this connection we need to recall from physics that entropy is the tendency of any closed system to go from higher to lower degrees of organization as it grows older, simply because the lower states of organization are more probable. Norbert Wiener puts it thus:

> As entropy increases, the universe, and all closed systems in the universe, tend naturally to deteriorate and lose their distinctiveness, to move from the least to the most probable state, from a state of organization and differentiation in which distinctions and forms exist, to a state of chaos and sameness. In Gibbs's universe order is least probable, chaos most probable. But while the universe as a whole, if indeed there is a whole universe, tends to run down, there are local enclaves whose direction seems opposed to that of the universe at large and in which there is a limited and temporary tendency for organization to increase. Life finds its home in some of these enclaves.(3)

Here we see man, and all living creatures for that matter, precariously positioned in a "local enclave," where the tendency is antientropic, or opposed to the general tendency of the rest of the universe. What we should be amazed at is not so much man's distinguished and dangerous position of eminence (for all life is that eminence) but that, through his *feelings* or *emotions,* he is instructed on this matter long before any such sophisticated ideas about the universe enter his head, and even if they never do. Chaos to him means dissolution, disorganization, and death; order means life and growth and fulfillment—so much so, as we shall see, that he is always tempted to achieve a premature order, an oder at all costs, because he loves the feelings that come with order and hates the feelings that come with chaos.

Whether life originated on this planet or whether it started from interstellar dust finding uniquely favorable conditions here, the important thing for us to notice is that once it appeared, life constituted a qualitative change in the contents of the universe, not just more of the same. We who are so easily impressed by size and quantity should try to imagine pictorially what the Second Law of Thermodynamics is saying; namely, that in all that we now call the physical universe, all those uncountable stars, all those billions upon billions of galaxies "out there," and in all those mountains and masses of inanimate matter "down here," even with their fine structures of crystalline formations—in all that, there is a *lower* degree of order, a *higher* degree of chaos, than there is in a single large organic molecule of the type that gave rise to life on this or any other planet. This picture presents enough of a change in our usual evaluation of the contents of the universe to tempt a religious person to say that, verily, in the beginning, God created order out of chaos, having first created chaos—but we have not as yet learned the purpose of talking in this way. Whatever way we do permit ourselves to talk about the most inclusive reality we can think of or imagine, we still have to say that *IT,* in order to produce life at all, had to overcome chaos, had to set itself in a direction opposed to chaos, and had to be able to continue in that direction by creating even more order, *even more precarious life,* in that local enclave!

III

So much for the sensory situation in man, already compounded by his feelings, his curiosity, his imagination, his love of some sensations and hatred of others, his fear of chaos, and his fascination by order. But how does it stand with the other animals? How can we even attempt a formula for animal knowledge at its widest possible application, unless we can decide how far down the evolutionary scale it might be legitimate to press analogies between human knowledge and all other kinds of animal knowledge? We need a set of minimal conditions, the very *least* that must be there, for worms as well as for philosophers and theologians, in order for something resembling knowledge to take place. There must be a nervous system or anything that functions as a nervous system, there must be a sensory situation, and there must be the ability to act—and a formula to connect all three. This formula would run somewhat as follows:

> Whenever there is an interaction between a nervous system and the nonnervous system in which it is situated, such that messages from the nonnervous system (light, sound, heat, pressure, chemical action) are encoded in the nervous system, and in such a manner that the animal's future performance can be either perpetuated or modified by its past performance, knowledge of some sort is taking place.

It has been thought until recently that only nervous systems of the vertebrate type were capable of functioning in this way, but now, thanks to developments in biology and cybernetics, a much less provincial view of the matter is coming into favor. Among invertebrates, that little darling of the biologists, the planarian worm, has shown itself capable not only of conditioned reflex behavior, but also of "instrumental learning" and of a type of behavior that resembles decision making, or "vicarious trial and error." (4)

In a more whimsical vein, Jerome Rothstein some years ago wrote what amounts to a hypothetical epistemology for a hypothetical worm, in an article called "Wiggleworm Physics." (5) Wiggleworm belongs to a "race of blind, deaf, highly intelligent worms living in black cold seabottom muck, and possessing only senses of touch, temperature, and a kind of taste (i.e., chemical sense)." The author then shows how science might develop for such creatures, using only such concepts as they are capable of developing by intelligent and imaginative thinking about the experience available to them. What would Wiggy's world look like?

> At any one time, wiggleworm science would probably not bear a close resemblance to our own. His notions of energy, for example, might arise from chemical thermodynamics rather than from mechanics. Energy might have "dimensions," not of mass, length, and time, but perhaps of force, volume, and temperature. What meaning, then, can one ascribe to the question as to whether Wiggy and we can come up with equivalent pictures of the universe? (5)

Rothstein is led to conclude that if Wiggy's criteria for the validation and improvement of his theories were the same as ours, and if he was always correcting them and filling them out to accord with new experience, when we had both covered the same accessible portions of the universe, Wiggy and we could end up with pictures of the same body of experience that were isomorphic. I only mention this epistemological fairy tale to show how, in relation to the actual planarian worm, and with considerably more information than we now have, we could construct a picture that would be dynamically isomorphic with ours of how the world must look to the worm.

But why stop with worms? It is certainly well known that even the most primitive single-celled animals have a directional center in or near the nucleus. Instead of getting into an argument over what is, or what is not, a nervous system, let us approach the matter from the more abstract viewpoint of cybernetics, the theory of messages. In the book already referred to, Norbert Wiener generalizes on some points that life-imitating machines (electronic brains) must have in common with nervous systems:

One is that they are machines to perform some definite task or tasks, and therefore must possess effector organs (analogous to arms and legs in human beings) with which such tasks can be performed. The second point is that they must be *en rapport* with the outer world by sense organs, such as photoelectric cells and thermometers, which not only tell them what the existing circumstances are, but enable them to record the performance or non-performance of their own tasks. This last function, as we have seen, is called *feedback,* the property of being able to adjust future conduct by past performance. Feedback may be as simple as that of the common reflex, or it may be a higher order feedback, in which past experience is used not only to regulate specific movements, but also whole policies of behavior. Such a policy-feedback may, and often does, appear to be what we know under one aspect as a conditioned reflex, and under another as learning. . . . Thus the nervous system and the automatic machines are fundamentally alike in that they are devices which make decisions on the basis of decisions they have made in the past.(6)

It is easy to see that such a cybernetical interpretation of how the nervous system operates roughly parallels our formula for animal knowledge at its widest application, and to apply the formula to the great variety of existing animals would mean to begin to study the varieties of input, encoding, decoding, output, feedback, further encoding, further decoding, modified output, modified feedback, and so on, and so forth. Just in the matter of output, think of the enormous variety in the repertory of responses available to the different species, ranging from extreme rigidity to considerable flexibility, permitting learning of different degrees all up and down the evolutionary animal scale. It is simply more obvious to us that, in the higher types of organisms,

The environment [i.e. the specific niche in the nonnervous system] considered as the past experience of the individual [i.e., as messages encoded in the nervous system] can modify the pattern of behavior into one which in some sense or other will deal more effectively with the future environment. In other words, the organism is not like the clockwork monad of Leibnitz with its pre-established harmony with the universe, but actually seeks a new equilibrium with the universe and its future contingencies. Its present is unlike its past and its future unlike its present. In the living organism as in the universe itself, exact repetition is absolutely impossible. (7)

If we now combine this broad view of knowledge with the cybernetic view of *life* (that it is a "local enclave" in which an antientropic tendency exists, opposed to the general tendency of the universe to run down to a state of disorganization, chaos, and sameness), and if we remember that messages themselves are a form of pattern and organization, having a "neg-entropy" of their own, we begin to glimpse a far-sweeping, even breathtaking, view of the role played by knowledge in the cosmic biography of the universe. At some point in the long, slow development of the chemical elements from the physical materials of the universal plasma, the antientropic tendency produced a molecule of a highly improbable degree of complexity and organization (something like the DNA or the RNA molecules) having the unique property of encoding messages from the less complex part of reality and using this "memory" to direct its own future actions in a direction favorable to nondestruction. That was the beginning of the whole life process we call the animal kingdom, with its almost infinite proliferation into various niches of the nonnervous system, a process that up to now has ended with the nervous system of man and the extension of its sense organs and some of its other functions through man-made, life-imitating machines.

Such a view seems to render all but incredible the time-honored problem known as the philosophical doubt of the existence of an objective, external world—or, the existence of that which I have called here the nonnervous system. All scientific evidence indicates that, in the order of cosmic development, the nonnervous system was here first, and that, when the time was ripe and it became possible for a part of reality to encode itself in another part of reality, the nonnervous system, even though more chaotic, forced or constrained the nervous system to correspond to itself in some degree, to establish a relationship that we might call *homeomorphic.* And whatever we may think of the evolutionary process morally, that is what it has accomplished for us neurally. All nervous systems that have survived up to the present time are homeomorphic with *some* part of the "external world," because in the long weeding-out process of evolution, all nervous systems that were not sufficiently homeomorphic with their niche in the nonnervous system were eliminated. They could not perform the job described in the formula. As we shall see, this also holds for man, and in a special, tragic sense, it holds in addition to and beyond the way it holds for the other animals. For man is the only animal that can, as it were, voluntarily, gratuitously, "wreck" his own nervous system, making it insufficiently homeomorphic with the situation in which it has to operate; and this he can do by virtue of one of the many possible ways of using his imagination, which is to feed back into the nervous system all sorts of lies and fantasies about himself and the world, instead of "signals" from the actual situation.

We now temporarily take our leave of the kingdom of the animals, to return again in Chapter Four. The formula for animal knowledge on page 27 applies of course, also to man; but this formula has had to be made so abstract, in order to cover the wide variety of nervous systems and sensory situations, that it is almost useless in helping us to see what happens in the case of man. (The built-in danger of abstractions!) We must now flesh out this formula with specific human details. Let us not, however, fall into the trap of trying to find a single outstanding characteristic of man which, beyond a shadow of doubt, distinguishes him from the other animals—by saying, for example, that man is preeminently the animal who uses language, who constructs artifacts, who reasons about the world, who plays games, who makes religions, or who questions himself, laughs, doubts, and so on. A list so easily multiplied patently defeats the notion of the single defining characteristic. Nor is it necessary to look for any "sharp cuts" between man and the higher vertebrates in order to insure him his ambiguously honorable status. Just anatomically, for good or ill, man has an overdeveloped brain, and whatever his brain can do in an overdeveloped degree, the others can do in an underdeveloped or rudimentary degree. It should be obvious, then, that every such overdeveloped activity will affect every part of the complex interaction described by the formula for all animal knowledge.

REFERENCES

1. Jerome S. Bruner, Jacqueline J. Goodnow, and George A. Austin, *A Study of Thinking,* John Wiley & Sons, Inc., New York, 1956, p. 1.
2. Philip Wheelwright, *Metaphor and Reality,* Indiana University Press, Bloomington, 1952, pp. 116 and 122.
3. Norbert Wiener, *The Human Use of Human Beings,* Houghton Mifflin Co., Boston, 1954, p. 12.
4. Jay Boyd Best, "Protopsychology," *Scientific American,* February 1963.
5. Jerome Rothstein, "Wiggleworm Physics," *Physics Today,* September 1962, p. 38.
6. Wiener, *The Human Use of Human Beings,* p. 33.
7. Wiener, *The Human Use of Human Beings,* p. 48.

Three Imaging, Imagining, and Vain Imaginations*

I

This chapter tries to clarify the functioning of that activity of the human type of brain which is now all too indiscriminately denoted by the term "imagination." For, as we have seen in the previous chapter, it was impossible to describe even sensory perception in man without noticing how this faculty was proliferated and expanded, especially in the case of vision, to form a world promising order and threatening chaos, by the extension of vision in imagination. In terms of the formula for all animal knowledge given in Chapter Two, imagination must be placed between the input and the output, since it influences not only the sheer quantitative size of the input but also the qualitative extent and variety in the repertory of responses, or the output.

The problem is to separate out and delineate, and if possible to distinguish by a descriptive name or phrase, at least a few of the more prominent overlapping, interlacing, often simultaneous operations of the mind that we now loosely band

* For the source of the term "Vain Imaginations" see Appendix Two, Section II.

together under the name "imagination." We want to see if in this way we cannot get a better idea of the part played by these distinct operations in the knowledge process, either to promote it or to hinder it.

Verbal entanglements due to the failure to distinguish the different functions of imagination have already turned much of our aesthetic and psychological discourse into a terminological jungle, but at least it is a problem of ancient lineage. Thus Aristotle uses the word *phantasiai* to stand for "mental images" and placing them in a position intermediate between sense perception and understanding. Sense perception is a prerequisite for mental images, and mental images, in turn, are a prerequisite for understanding. The mental images, Aristotle seems to mean here, isolate, perpetuate and juxtapose the sense perceptions before the mind in a way that gives the intellect objects of contemplation for the further purposes of comparison and contrast, classification, and concept formation (i.e., "understanding"). On the other hand, in his treatise *On the Sublime,* Longinus uses the same word, *phantasiai,* to stand for "vivid poetic imagery," recommending that the would-be great writer make use of this particular device, among many others, in order to "excite his audience" and in order to "master the hearer rather than persuade him." (1) The very purposes so unabashedly set before the rhetorician here apparently exclude the possibility that Longinus was using the term *phantasiai* in any sense similar to Aristotle's *phantasiai,* which are aids to intellectual understanding. How nice it would be to have two different words; and yet we are not much better off in English when blithely we use the word "imagination" in both senses.

As a first approximation, the chief difference between Aristotle's "mental images" and what Longinus calls "vivid poetic images" (both using the word *phantasiai*) seems to be that the former are not motivated by any creative, distortive, manipulative or emotion-evoking purposes, whereas the latter are just so motivated. Leaving these for the next section, let us concentrate on the first, and see if we cannot retain as much as possible of Aristotle's meaning without walking into the verbal trap invitingly held open by the term "mental images." This innocent term inevitably implies a contrast with "physical images." Yet if we ask ourselves what are physical images and where are they to be found, we see that there are only objects and images, so that when you have said the word "image," you have already, in this context, said the word "mental." (Of course, in another context, you might say, the sculptor made an image of the animal in stone, meaning a created object.) For this reason, and also because we are interested in operations rather than in attributes, let us change Aristotle's "mental images" into a more active form: *imaging.* And in order to indicate the absence of motivation for any creative, manipulative, distortive, or emotion-evoking ends, let us further call this function of the mind *imaging-as-if-perceiving.*

Here we note again the dominance of vision in the human sensory situation, in the great poverty of words for describing the activity of imaging-as-if-perceiving. We have to use the single word "image," which refers to the sense of sight, for all the other senses as well as sight. Thus we have to talk about auditory imaging, tactile imaging, the imaging of tastes and smells, and the imaging of internal sensations in the body, such as visceral and kinaesthetic sensations. This awkwardness of terminology does not mean that the act cannot be performed—and with the greatest of ease. Thus if I ask the reader to image-as-if-perceiving the taste of strawberries, the aroma of coffee, or the sound of a clarinet, he will have no difficulty in doing this; but he will probably start off the specific sensory modality in each case by means of a visual image of the object in question. Not only that, but the mere mention of the object-word will probably bring along with it a train of memory images in connection with the object as perceived in past experience, including the greatest perceptual details, such as the taste of strawberries with and without cream, and the taste of unripe and overripe strawberries. If I were then to ask the reader to describe what he can so easily image-as-if-perceiving, there would be embarrassment all around at the paucity of words.

Imaging-as-if-perceiving can take place both in the presence of the object and in its absence. When it takes place in the presence of the object, it has the effect of putting the subject in the privileged position of a perceiver who is able to see things from many different angles simultaneously. Thus when I simply look about me, I do not perceive; but I can easily image-as-if-perceiving the far sides of objects, the undersides of tables, the insides of cabinets, objects situated in back of other objects, moving objects passing behind other objects. There is thus created a feeling, not a concept, of the three-dimensionality of space, of its being a "room" for objects to occupy and move about in, and of the subject himself moving about in it with his godlike *super-vision* of its simultaneous contents. For a grownup person, the radius of this "room" is further extended by means of imaging-as-if-perceiving in the absence of the object, and, as we noted before, there is really no limit to the size of this "room" except the sheer weariness of the job of attempting to image-as-if-perceiving the absent objects and the spaces between them. The *concept* of space, as a measurable extension, comes, as it were, to the rescue of this perceiving combined with imaging-as-if-perceiving, to pace off the larger distances with giant steps—miles, light-years, even—and thus to save us the labor of calling up images. And labor it is, as we can see by realizing that we take the trouble to summon images accurately and completely only when some practical activity depends for its success on our having correct images in our head beyond what we immediately perceive—such as when we are driving a car, going on a trip, performing any dangerous task.

Imaging-as-if-perceiving takes place in relation to time as well as space, and here it is even more of a "labor"; for we are such visually oriented animals that we cannot "calibrate" the passage of time without filling it with a succession of visualizable events, happenings. Let the reader try to image-as-if-perceiving a very long period of time, say 100,000 years, in which "nothing" happens. Nevertheless, we all do it for short periods of time, especially for that privileged period we call our own biography, which we can "calibrate" with the visual images of our own remembered experiences. Imaging-as-if-perceiving in relation to time has the effect of creating a depth dimension in time, making it into another *lebensraum,* a "room" in which to move around, backward and forward, in past-present-future. As Proust remarked in *The Remembrance of Things Past,* every individual "drags around" with him a depth of "felt time" that corresponds roughly to the length of conceptual time through which he has lived, until finally this burden becomes too much for his failing energies and at the end he is content to lay it down. But of course no (educated) grownup person imagines that this "lebensraum" of time begins and ends with his own birth and anticipated death. The function of education is to stretch out and furnish with as rich a content as possible the time-room as well as the space-room. Historians and archaeologists, for example, acquire such a virtuosity in imaging-as-if-perceiving past eras with the assistance of historical data, that a given era becomes an adjunct—a province, as it were—of the era in which they themselves are living. All learnable history, for those who wish to take the trouble of doing the imaging-work, can in this way be added to each individual's biography as its preliminary and preparatory stage, creating a "life-span" of incredible length and eventfulness—in comparison with which the ordinary life-span of those who are not interested in history seems as deprived and abbreviated as if they were suffering from extensive areas of amnesia.

Taken together, the spatial and the temporal imaging-as-if-perceiving alone create around each individual the beginnings of a space-time *world,* radiating from him at the center in all directions in space, and existing before and after him from the "now" of his time. Let us call the creation of this space-time world by means of the extrapolation of perception the *matrix imaging,* and the world thus created the *matrix world.* The *matrix world* in its original primitiveness, as in children, is devoid of concepts—it is what must be there first in order that the comparisons, contrasts, and classifications that lead to concept-formation can be made. Thus it makes no sense to talk about whether the primitive *matrix world* is subjective or objective, public or private, "true or false," for these are discriminations and distinctions that can be made only after the matrix world has supplied the raw materials for making them.

Although the matrix world in its beginnings is devoid of concepts, it is not devoid of feelings—and these feelings are examined in greater detail in Chapter Four. Here we need only note that they are just the feelings produced by certain sensory perceptions in the percipient and reproduced faithfully by imaging-as-if-perceiving along with the images. The imaging-as-if-perceiving of terrifying objects reproduces the terror along with the objects; applied to pleasing objects, the process reproduces the pleasure along with the objects. It is doubtful whether this should be called the "law of association," for this law implies a much more mechanical, extrinsic relation between the two, as if the image of the object somehow "dragged in" the feeling "associated" with the object. It would seem, rather, that the object is perceived as that which produces the feeling in the percipient, and the imaging-as-if-perceiving reproduces this phenomenon. A recent study has shown that prenatal children respond to music that is transmitted into the womb through the mother's abdominal wall, the response being a differential acceleration of the prenatal child's heart beat. Surely it is an awkward way of putting it to say that the unborn child experiences this feeling of increased heart beat because he has it "associated" with certain musical sounds.

The *matrix world* both in its image and feeling content, and later even more in its conceptual content, differs very widely, as between children and adults; among adults it differs as between educated and uneducated persons; and among educated persons regarding the direction in which attention is focused for the expansion of perceiving through imaging-as-if-perceiving. According to the formula we set up for animal knowledge at its widest application, including all the other animals, the *matrix world* should be regarded as the *locus,* the place of interaction between the nervous system and the non-nervous system in its most primary or elementary form. Let us use the term *matrix world* to distinguish the human type of consciousness from the animal kinds. Because for man, even the most primitive form of interaction, thanks to the imaging activity, creates a world quite different from that of the animals. Certainly the animals must be capable of some degree of imaging—cats and dogs learn to come "home" from places from which they can not perceive their destination—but it is precisely the "overdevelopment" of this activity in man that makes the difference. Cats and dogs do not, even as "grown-ups," live in a world that existed before they were born, and will continue to exist after they die; nor do they exist in a world of overwhelming spatial extension and particular content.

Furthermore, it is from this matrix world that man receives his earliest "feedback" signals. What we call the natural hedonism of the young child is nothing but his learning from the "feedback" about the success or failure of his earliest "moves"—success being pleasure-producing, failure pain-producing. That is why the feeling side of the imaging is so important for children, absolutely essential to survival (although we see later that the feeling side in a much modified and developed form is quite as important for the adult). People who deplore the unbridled hedonism of small children should remember that the most intractable form of insanity is what used to be called *dementia praecox,* an *apathy* of such totality, profoundness, and impenetrability that it must be credited to some defect in the nervous system. Even among the "normal," there is nothing quite so pathetic as an apathetic child. This fact also brings out the *interaction* aspect of knowledge, for unless the child makes a *move* toward what seems to be the pleasurable object, he never will get the feedback that educates his future moves. Of course he also needs the protective "feedback" from the parents to make up for his limited perceptions and for his lack of conceptual thinking at this stage.

Now this *matrix imaging,* or imaging-as-if-perceiving in space and time, must be contrasted with a far more prevalent and omnipresent form of imaging, which is mostly of a random, diffuse, and fleeting character and which we may call *impressionistic imaging.* This is the kind of imaging that accompanies, in various degrees, all reading, thinking, feeling, desiring, daydreaming, reminiscing, meditating, interior monologue, and vague planning. It, along with its feelings, is usually *interwoven* with words: the words calling up images and feelings, the images and feelings calling up words, in the kind of continuous flow that writers try to capture with the "stream of consciousness" style. (We must, of course, assume that the preverbal child also has a stream of consciousness, without the verbal interlacing.)

In impressionistic imaging there is very little attempt at assuring correctness of the image, because there is no time for that, the images come and go so instantaneously. Sometimes the images are the faintest traces, other times they are vivid; sometimes they are like single flashes, other times in whole clusters or trains. Some of the images are perfectly neutral in their charge of feeling, others are strongly loaded, even capable of activating an emotional upheaval. Some people claim they can think without any imaging at all, using words only. No doubt this is possible, but it means they are using words as a logical shorthand, subtracting from the individual word everything except its logical and definitional relations to other words. This probably makes for speedy thinking, but the alleged nonimagers forget that they learned the meaning of the word in the first place with the assistance of images (unless it is strictly a technical term, defined, e.g., as, "let us call relation X by the term Y").

It is amazing with what effortlessness and silence (no grinding of gears!) we can switch back and forth between these two types of imaging. We are reading a novel, say, and we just allow the succession of words to evoke in us a spontaneous flow of images, perhaps mixed with other words and the feelings corresponding to them—when all of a sudden we come upon a scene that the author makes us feel is important—a town, say, complete with moods and weather conditions. Instantly we switch from impressionistic imaging to imaging-as-if-perceiving, using every word the author gives us as a clue to the correct image until we see the very town before us and feel the very feelings the author wished it to convey. But imaging-as-if-perceiving is a lot of work, and as soon as the narrative becomes less interesting we switch back to the impressionistic type of imaging. That is why we feel disappointed in a novel that does not reward us proportionately to the amount of imaging-as-if-perceiving work we have invested in it. The same switching back and forth holds for the reading of scientific material. For most people, reading the description of scientific data and the development of a theory from the data, no matter how "conceptual" the argument may become, is accompanied by a certain amount of impressionistic imaging; but as soon as an apparatus is described, or the parts of an animal, there is an instant switching to imaging-as-if-perceiving. In modern textbooks, however, there is a strong tendency for the author to save himself a lot of words and his readers a lot of imaging-as-if-perceiving work by printing in the textbook a diagram or a photograph.

The title of this chapter, "Imaging, Imagining, and Vain Imaginations," has suggested both the possibility and the desirability of distinguishing several different activities of the mind within the general category of the imagination, and already we have been able to distinguish two kinds of imaging: imaging-as-if-perceiving and impressionistic imaging. These can be thought of as the "raw materials" for a more constructive, inventive, manipulative operation (called "imagining" in the next section). But just as they are, in their passively faithful reproduction of sense perception from memory or association or from the implications of conceptual knowledge about how things must be where they cannot be immediately perceived, they form a large part of the operations of consciousness in its many-leveled simultaneity. And, as we see in the next chapter, they are what makes possible the simplest kind of conceptual knowledge, since they present to the operations of the intellect, which are concerned primarily with relationships and classifications, definite images, isolated, perpetuated, and juxtaposed, so that they can be classified and their relationships can be observed (Aristotle's "mental images").

II

We come now to those operations of the mind I propose to call *imagining,* whose chief distinguishing mark from *imaging* is that to some degree there is always present an active, purposive, manipulative, desire-gratifying, or future-rearranging element. If I am sitting in my room in the dead of winter gazing out at my snow-covered garden and trying to think how it will look at the end of March when it comes time to plant certain shrubs, I am imaging-as-if-perceiving my garden. But if in addition to this I try to think how the garden will look after I plant a rhododendron here and a row of rose bushes there, I am *imagining* my garden. This is a very primitive example, but it shows up at once one of the immediate advantages of *imagining*: how easy it is, in imagining, to move those shrubs around to try them out in different places, as against hauling the actual bushes first here and then there, to see where they look best! It has well been said that highly imaginative persons are physically lazy: they try to save themselves work by performing all acts in imagination first, to see which way is easiest, and indeed to decide whether the whole thing is worth doing at all.

The object of this little "anatomy of the imagination" is certainly not to multiply entities but to distinguish the least number of its functions whose dissection and separation might be useful in the avoidance of verbal confusions. Therefore, I describe only three types of imagining, *projective imagining, empathic imagining,* and *expressive imagining.* Distinguishing them from one another seems to me to be justifiable because, although certainly they all may be present simultaneously in any special occasion of imagining, they are concerned with, or directed toward, different objects or goals. *Projective imagining* is always directed toward the future; *empathic imagining* is always directed toward the inner life of other people, or one's own at other times, or some imagined person's, or even some animal's; and *expressive imagining* is always concerned with the desire for, and the problem of, communication, even if only to "talk to oneself."

Projective imagining, as the name indicates, always involves a *project,* either a thing or an action to be brought into being in the future—something that has not existed or happened up to now but is being contemplated as a possibility and considered for realization. Even if the subject of the project is entirely concerned with the past, such as digging up archaeological remains or reconstructing a geological era, it is a project because it has to be imagined first as something that will be done in the future. The little example about the garden is an instance of projective imagining. It involves almost no element of the other two kinds, except perhaps as one imagines, empathically, one's own future satisfaction in having improved the plantings around the house.

Man's life is one project after another, and many simultaneous projects; thus it is hard to know where to start in giving examples of projective imagining. It is difficult to name a single activity of everyday life, from preparing food to doing one's job to raising children, that does not involve projective imagining to some degree. Of course, many of these projects become so habitual that one no longer notices the work of imagining that made them possible in the first place, except when it becomes necessary to "correct" or "adjust" them to changing circumstances. Habit is a great labor saver, and the work of projective imagining is a great part of the labor that it saves.

Projective imagining involves putting together or somehow rearranging things or actions in some new combination that is desired for some particular end, hoping to avoid unnecessary moves, mistakes, dangers, and various forms of failure by "fore-seeing" them. Hence projective imagining as such cannot be thought of as intrinsically constructive or creative, since all destructive projects, such as wars and murders and robberies and "organized" crimes, precisely to the extent that they are "pre-meditated," involve just as much projective imagining as constructive and creative projects (perhaps even more, because they are likely to encounter more resistance than cooperation). We have only to think of the amount of projective imagining that goes into the making and delivering of the modern weapons of war. Nevertheless, in spite of its use in destructive projects, we tend on the whole to think of projective imagining in constructive terms, because the entire visible structure of our civilization and culture bears witness to the enormous amount of projective imagining that man has expended on its creation and continues to spend on its change and growth. Here, in the capacity for projective imagining, is the ground on which the *"homo faber"* school would like to found the chief distinction between man and the other animals; but it is only one, and perhaps the simplest, form of imagining. Why don't the animals construct things? Except on an inherited pattern, as when the bird makes a nest or when the ape uses the stick to get the banana, animals do not seem to have even the desire to make anything. At least, we do not see them trying and failing, as we see little children trying and failing to make something with their toys. Apparently it comes down to a matter of attention and concentration: being able to hold the imaged items together in the new arrangement long enough and clearly enough to see the desirability of the imagined combination. Or why, we might ask, don't the animals carry out "plans of action"? As far as we can see, they do not, although here it would be more difficult to distinguish clearly between planned and unplanned actions. Projective imagining seems to be what is missing in their mental makeup, even though many are quite intelligent and can be taught specific kinds of behavior. Observing a "seeing-eye" dog at work, we suspect that the dog must be using some kind of projective imagining in a rudimentary form.

Although projective imagining plays its most obvious and outwardly visible role in practical life (i.e., in constructed things and planned actions), it is constantly at work in the background of every kind of thinking, if only to apply the negative, to veto any budding projects that may arise. One has only to remember a few of the projects he has thought of and never carried out, because, thanks to projective imagining, they were found to be unfeasible, undesirable, or not worth the effort. There is a distinct danger that too much projective imagining, applied too assiduously to all prospective acts, can deprive human life of all spontaneity and impulsive response—a danger we see in the contrast between the delightful "thoughtlessness" of children at play and the restrained, calculated behavior of adults in social groups. And yet "thoughtlessness" in a grownup can easily become less than charming, for the exercise of projective imagining—combined with the other two, empathic and expressive imagining—is a prerequisite for all civilized behavior, all ethical directives concerning how people should be treated, all person-relations that transcend thing-relations. At first, projective imagining warns us against evil things we ought not to do because of the consequences the actions will bring on ourselves. In the long run, however, by providing an image of what human life in society would be like if everyone did such acts, projective imagining leads us not to want to do them at all—the germ of the "do not do unto others what you would not have them do to you" dictum.

Projective imagining plays its part not only in the making of civilization but also in the creation of culture, for every item of cultural creation is in some degree also a constructed thing or a planned action or both. For example, a scientific hypothesis explaining a natural phenomenon is always to some degree both a constructed thing and a plan of action; therefore it involves projective imagining on both counts. It is a "construct" of concepts, using perhaps even some visual "models," in which some new and previously unsuspected connections and relationships are proposed, and it is a plan of action whereby the correctness of the proposed connections and relationships may be experimentally verified. The actual results of the action, the laboratory findings, suggest to the theorizer how the hypothesis may have to be modified to accord with them, or perhaps how a crucial experiment could be devised to decide between two conflicting hypotheses—sometimes a veritable feat of projective imagining, since only a few of the variables involved are under laboratory control.

In this way, by the reciprocal interaction of hypothesis and testing, the huge body of self-consistent scientific thought is gradually built up; but there is nothing automatically "self-correcting" about this "scientific method," as is commonly thought. Unless projective imagining produces ever new constructs of concepts and plans of action—the former proposing, the latter disposing—there will not be anything to correct, and the theory will not grow, will not expand, or harmonize with other theories. The character of projective imagining in science varies enormously, with the variety of problems and with the different types of scientific minds. It ranges all the way from the systematic ransacking of many possibilities simultaneously by huge, subsidized teams of researchers (as in the project to produce the atom bomb or in the "war on cancer") to the lonely, patient, preoccupation with a single problem by the solitary thinker, sometimes illuminated by sudden flashes of insight, when it is as if two previously unrelated Gestalts in the mind suddenly meshed to provide an unexpected solution—an intellectual occurrence described by Einstein and Poincaré, among others.

When we turn from the sciences to the arts, and especially to the creative arts as opposed to those concerned with the acquisition of erudition or techniques, projective imagining tends to become overshadowed by the presence and prominence of the other two, empathic imagining and expressive imagining. But it is there nonetheless, if any *work* of art, and not merely daydreaming, is to be the issue. Every such work, in whatever medium, is to some degree a constructed thing or a plan of action or both; and to that extent each involves some projective imagining. Let me give now a preliminary definition (brief description) of art, and we can run through a few examples just to see how projective imagining is involved in them.

Art, let us say then, is the embodiment or enactment or reenactment of experience as clarified by imagination, in a limited sensory modality. "Embodiment" obviously must issue in a constructed thing, and "enactment" in a planned action, and the "sensory modality" will partly determine what definite form the project will take, even though, all along, empathic and expressive imaginings are exercising the greatest influence over it, trying to make it go this way rather than that way, to avoid this, and to achieve that. Not every experience can be embodied or enacted in every sensory modality—a great blessing to the artist—for this limitation, along with his own sensory limitations, helps him to arrive at a definite, completed *work*. Otherwise his creativity would be dispersed in all directions, wandering about and wearing itself out in a formless cloud of infinite possibilites. And even so, if we simply list the major sensory modalities used by art for its embodiments or enactments—music, painting, sculpture, architecture, poetry, fiction, drama, dance, and film—and if we somewhat playfully "multiply" these by the number of artists working in them, and then again "multiply" by the number of different experiences seeking embodiment or enactment in each kind, we are soon made dizzy by the astounding variety of possibilities presented. This is the famous inexhaustibility of art, guaranteed to provide endless employment for projective imagining. Music, for example, may be regarded as an enactment of certain kinds of life-feeling in a formal structure of sound, involving at least a definite composition and plans for specific instruments to play it. Sculpture may be regarded as the embodiment of other life-feelings into free-standing volumes, involving much more obviously a constructed three-dimensional thing, much less obviously the planned action of walking around it to see it from all sides. I leave it to the reader (as a kind of homework) to puzzle out the others, to see just how each of them, in specific instances, involves some degree of projective imagining.

III

Empathic imagining and *expressive imagining* should really be considered together, because in many cases they are so intimately and reciprocally related. Not all empathic imagining issues in expression, but if it does, then we may say, approximately, that empathic imagining is concerned with the "what" of the matter and expressive imagining is concerned with the "how" of its clarification, formulation, and communication. Everyone knows from experience how these two forms of imagining reciprocally urge each other on—how a first attempt at expression, being found inadequate, leads to a more thorough empathic imagining of the situation, followed by an attempt at a better expression of it, and so on.

Let us begin with empathic imagining and the question of how the term "feeling" should be used in this context, to indicate the "pathic" part of the word empathic. We noted that empathic imagining is directed toward the interior life of other people, or one's own at times other than the present, or some imagined person's, or even some animal's. The word empathic tries to parallel the German word *Einfühlung,* a kind of imaginative, feeling-one's-way-into something otherwise strange or closed to perception. It carries less of the connotation of "sympathetic understanding" than does the German word, however. To cover the range of "interior life," the English word "feeling" seems too broad and the word "emotion," as defined by Webster, seems too narrow.

Under the broad, less specific word "feeling" we find:

> *3a:* the undifferentiated background of one's awareness considered apart from any identifiable sensation, perception, or thought. *3b:* the overall quality of one's awareness esp. as measured along a pleasantness/unpleasantness continuum.

Under the narrow, more specific word "emotion" we find:

> *1c:* a physiological departure from homeostasis that is subjectively experienced in strong feeling (as of love, hate, desire, or fear) and manifests itself in neuromuscular, respiratory, cardiovascular, hormonal, and other bodily changes preparatory to overt acts which may or may not be performed. *1d:* an instance of such a turmoil or agitation in feeling or sensibility: state of strong feeling (as of fear, anger, disgust, grief, joy, or surprise). (2)

The difficulty seems to be that, for *qualitative* description, "feeling" is inherently too vague, while "emotion" describes *strength* of feeling, in which physiological turmoil and agitation draw the attention away from the qualitative characteristics, except in the most general terms.

For a more qualitative distinction between feeling (whether strong or mild) and sensation, we turn to Otto Baensch's description of "objective feelings" in his essay "Art and Feeling." (3) Objective feelings (as in a "sad" face, a "moody" landscape, or a "gay" rug), Baensch says, are always embedded and inherent in objects from which they cannot be actually separated, but only distinguished by abstraction. They are best described as nonsensory qualities somehow inherent in sensory objects. They are distinguishable from the sensory qualities because

> [t]he latter stand in relation to each other, they are combined, and composed so as to produce jointly the appearance of the object. Non-sensory qualities, on the other hand, surround and permeate this whole structure in fluid pervasion, and cannot be brought into any explicit correlation with its component elements. They are contained in the sensory qualities as well as in the formal aspects; and despite all their own variety and contrasts, they melt and mingle in a total impression which is hard to analyze. (4)

Subjective feelings (i.e., the feelings the subject himself brings into the presence of the object), Baensch goes on to say, are also nonsensory qualities, but they are embedded or pervasive in the images, memories, anticipations, purposes, thoughts, and valuations whose perpetual flow constitutes perceptions or somatic changes—as is done in the definition of "emotion" above—simply stating that, no matter how interwoven in our psychophysical nature these two might be, they are still two different things. (Here Baensch does not consider that the "strength" of the feelings might determine the degree of somatic involvement.) The subjective feelings, being qualitatively similar, flow back and forth, mingle with and mutually modify, and are modified by, the objective feelings, so that in the end we do get something akin to the quality of "consciousness itself" in the broad sense of the word "feeling." The subjective feelings are also recalcitrant to conceptual treatment "as soon as we try to go beyond the crudest designations; there is no systematic scheme that is subtle enough in its logical operations to capture and convey their properties." Furthermore, Baensch regards the designation: pleasant/unpleasant as descriptive of the person's desire to retain or get rid of certain feelings, rather than descriptive of the specific quality of the feelings themselves.

> We are really confronted with a chaotic manifold from which no internal principle of order can be derived. Nothing therefore avails us in life and in scientific thought but to approach them indirectly, correlating them with the describable events, inside or outside ourselves, that contain and thus convey them; in the hope that anyone reminded of such events will thus be led somehow to experience the emotive qualities, too, that we wish to bring to his attention." (5)

Baensch, of course, argues in the remainder of his essay that art, primarily, is best fitted to approach the feelings in the indirect manner just described; that is to say,

> [A]rt, like science, is a mental activity whereby we bring certain contents of the world into the realm of objectively valid cognition; and that furthermore, it is the particular office of art to do this with the world's emotional content. According to this, therefore, the function of art is not to give the percipient any kind of pleasure, however noble, but to acquaint him with something which he has not known before. (6)

This is the basic thesis of all cognitive theories of art.

R. G. Collingwood, another cognitivist in art, suggests that probably every individual sensum carries what he calls, somewhat physically, an "emotional charge," but that it would be hard to prove this because we are ordinarily aware of great masses of sensa simultaneously and furthermore have been conditioned to pay more attention to the sensa in their specific patterns than to the emotional charges on them. (Collingwood calls the latter a habit "sterilizing the sensa.")

In persons who are likely to read this book, the habit of sterilizing sensa has probably become so ingrained that a reader who tries to go behind it will find it very hard to overcome the resistance which hampers him at every move in his inquiry. In so far as he succeeds in recognizing what really happens in himself, I believe he will find that every sensum presents itself to him bearing a specific emotional charge, and that sensation and emotion, thus related, are twin elements in every experience of feeling. In children this is clearer than in adults, because they have not yet been educated into the conventions of the society into which they have been born; in artists clearer than in other adults, because in order to be artists they must train themselves in that particular to resist these conventions. (7)

I think that Baensch, although he might have approved the phrase "emotional charge," would have disagreed with Collingwood's rather atomistic linkage of emotion-sensation, since Baensch pointed out that it is the whole composite Gestalt of sensations we call an object that carries the "emotional charge" pervasively, not, as it were, in little bits, related and composed as the bits of sensa are.

The most consistent and thoroughgoing cognitivist in the theory of art is Susanne K. Langer, who in *Feeling and Form,* expanded her cognitivist theory of music to cover all the other arts, each specifically investigated to determine what it is trying to do with its own sensory materials. In music, her cognitive thesis rests on the observation that "music is a tonal analogue of emotive life":

The tonal structures we call "music" bear a close logical similarity to the forms of human feeling—forms of growth and attenuation, flowing and stowing, conflict and resolution, speed, arrest, terrific excitement, calm, or subtle activation and dreamy lapses—not joy and sorrow perhaps, but the poignancy of either and both—the greatness and brevity and eternal passing of everything vitally felt. Such is the pattern, or logical form, of sentience; and the pattern of music is that same form worked out in pure, measured sound and silence. Music is a tonal analogue of emotive life. (8)

Excellent though her analysis of the individual arts is, she presents neither a psychological theory of even the simplest emotions or a general theory of the imagination, to connect her artistic cognition of feeling to other, that is, nonartistic, cognition of feelings and their function in life. Like all cognitivists in art, she has a very low opinion of psychology, probably because it deals with what Baensch calls "only the crudest designations." But this neglect does not tell us why it is necessary to get *beyond* these crude designations, and how art does get beyond them. Thus Langer leaves us with art as a rather aristocratic, rarefied pursuit, and with the question, who needs it? Which is no small question, in a technological era. Quite understandably, all cognitivists in art are also very *suspicious* of psychology, because of the utter travesty of the arts that has been committed by the Freudian interpreters. According to these writers, the artist is a kind of socially tolerated neurotic or even psychotic, who, through the accident of great talent, has been able to convert his repressed, frustrated desires and wish-fulfillments into public, marketable products (marketable precisely because the public shares these repressed desires but has no other way to gratify them). But more about that under "vain imaginations."

Empathic imagining in the everyday, nonartistic context is a kind of "educated guessing" at what is going on in other people's conscious states, guided by clues they give us. Since we tend to forget how much "education" has gone into the "guessing," it is real shock to recall that, in terms of the broad formula for knowledge as an interaction between a nervous system and its niche in the nonnervous system, other people belong to our niche in the nonnervous system (as indeed we belong to theirs), and the signals we receive from them, as species of signals, differ in no respect from those we receive from inanimate objects—they include sights, sounds, touches, tastes, smells, temperatures, and pressures. How is it, then, that we come to make such a clear distinction between human and nonhuman objects? Or, to put this question in its traditional philosophical form, how do we know that other minds exist? Surely it is from the nature of the responses we get from people with whom we are trying to communicate—responses that betray that *others* are practising empathic imagining on *us,* constantly correcting or changing their verbal (or facial) expressions to indicate that they are trying to understand what we say in a much larger context of consciousness than our words themselves convey. Where there is the will to communicate and plenty of empathic imagining, each party in a conversation "fills in" for the other party what one did *not* succeed in putting into words, constantly adjusting and correcting this filled-in part as he receives new clues that enable him to "follow" the other's thought and feeling. We recognize quickly enough whether this "filling in" has been done correctly for us by someone else. Sometimes we get a jolt, when as a result of new factual information, we realize that we have completely "misread" someone else's feelings, that is, made a blunder. Well, this is not the way we communicate with talking dolls or mindless creatures. We know other minds exist because we send them clues that only a mind could understand, and they send us back clues that only a mind that understood our clues could have sent. All this is dramatized to the point of comedy when a shared language is missing or in a game like charades. The language of facial expression, gesture, posture, acting-out, and so on, is something that we learn from early childhood both to use and to read, and to refine, as increasingly we discover that this language can be used not only for disclosing what we feel, but also for concealing what we wish not to reveal to others. Empathic imagining in communication between grownups can become a feat of virtuosity when we try to learn each person's own *style,* with its precise mixture of reserve and disclosure, of looking at us or looking away, of tonal variations, degrees of acting or pantomine, characteristic attitudes, masks of politeness, clues for showing or hiding surprise, annoyance, embarrassment—all this in *addition* to the words that are presumed understood.

It is scarcely necessary to point out the large part played by empathic imagining in regard to our expectations of the future, both in situations we willingly enter and in those which are forced on us by our circumstances. We say, for example, that we "dread" an interview; but what we dread are feelings of inadequacy, of embarrassment at making a poor impression, or of disappointment, which we empathically imagine we will undergo at that time. Sometimes we are quite mistaken and are pleasantly surprised—and thus, gradually, we learn to discipline our imagination by past experience and by the factual probabilities of future situations. In conjunction with projective imagining, we try to imagine empathically what precise satisfactions we will achieve by means of the constructed thing or the planned action. Here again, only experience will teach us not to over- or underestimate them in judging the worthwhileness of the project. However, if it is at all a long-term project, there is the mysterious process of our own becoming, of our own supposedly well-known feelings changing during and because of actually carrying out the project, of discoveries and disenchantments that could not have been made available to us in any other way—in the course of empathic imagining with regard to the future, these unpredictables that limit the "lesson-learning" from past experience are felt as challenges and temptations, even as irrationalities that threaten to reduce at least part of our goal-striving to absurdity. "If I had known then how I would feel now, I would have never. . . ." Yet this recurring experience does not stop our trying to anticipate the quality of feelings on future occasions, whether to seek or avoid them.

It is not nearly as obvious that empathic imagining also plays a part in our remembrance of our feelings and emotions in past situations, for here we can draw on the supply of recallable perceptions we call our memory, and it would seem that imaging-as-if-perceiving the past events would be sufficient to remind us of their emotive content. Yet we all know that remembered images are more reliable than remembered feelings, even though the latter may be "stronger," in the sense of being the real reason the images left their mark at all. The images can be recovered with considerable correctness by an effort of concentration; but the feelings they bring with them are as if seen through a telescope of their pastness, through a deep overlay of subsequent feelings. The images can be juxtaposed, side by side, regardless of temporal sequence, and viewed panoramically, but the feelings are such as we could have had only at a certain time in our personal history. Accordingly, the recollection of these feelings lends a temporal, sequential "depth" dimension to our own biography, as well as a sense of continuity and identity to our "self," for however strange they may seem to us now, we cannot doubt that they were once ours and not someone else's. The very nature of feelings as against images contributes to the difficulty of remembering them, for if, as Baensch wrote, they were a "chaotic manifold" at the time we experienced them, they are, as recollected, not easily identified (or, if clarified now, they seem to be changed by the act of remembering and thinking about them). Thus if poetry is "emotion recollected in tranquility," it is emotion modified by the act of recollection. Even if we were very thoughtful persons at the time being remembered and tried to "understand" our own feelings at that time, we know that the disturbing, "commotion" quality of feelings makes description and analysis not only difficult but tending to change or destroy the feeling itself that is being pinpointed for cognition. To quote Baensch again: "Thus feelings remain in shadow for our cognition, perhaps most of all at the time when our soul is completely suffused by them. Their fusion with the self keeps them arrested in the sphere of immediate experience; there they develop their greatest strength. We feel them, but we hardly see them." The effort to remember our past correctly is, after all, not so different from the effort to understand the interior lives of other people: it is a reconstruction from clues—in this case from the clues given by imaging-as-if-perceiving the past situation—plus empathic imagining of our feeling selves as they were at that time, not as we are now remembering them.

To make matters worse, or at any rate, more complicated, we read in the psychology books that memory tends to preserve the pleasant and to repress the unpleasant experiences of the past. If this were true, literally, everybody's recollected biography would be a veritable fairy tale. But it is not, as we know both from our own recollections and from those told to the psychoanalyst. Although the latter has to use "techniques," such as word-association and dream-association, to bring back the memory of extraordinarily unpleasant, emotionally charged, experiences, this does not mean that the rest of the biography remembered is extraordinarily pleasant, as any casebook will show. The average case history is a tale of woe, of failure and frustration, reluctantly brought forth under the sympathetic encouragement of the therapist. We may not *enjoy* remembering certain unpleasant feelings of past situations, but we can do it, so well indeed that the very thought of them causes us to reexperience the anger, anxiety, or helplessness that characterised them in the first place; thus the tendency not to dwell on unpleasant feelings may be merely a practical economy of the emotional life—not to add to present burdens past ones about which nothing can be done. This "economical" way of looking at feelings does seem to lend credence to Baensch's contention that the categorization of pleasant/unpleasant in relation to them characterizes less the real quality of the feelings themselves than our desire to retain them or get rid of them.

Because of the problem of the pleasantness or unpleasantness of feelings, and even more of strong feelings or emotions,

I have chosen the fairly neutral term *empathic imagining* for the operation of the imagination required to reconstruct them, either from our own past or from the interior lives of other people, according to the clues we are given or can find. I could have used the more popular term "sympathetic imagination." Obviously, however, it takes as much if not more effort of empathic imagining to participate vicariously in hostile or unpleasant emotions, especially when these are directed against us, as in pleasant ones, the very unpleasantness tending to hinder our doing an accurate job of full imaging-as-if-perceiving and paying attention to all the clues. Thus the fallibility or accuracy of empathic imagining is influenced both by our present attitude to ourselves as suffering (regardless of whether we are willing to relive an unpleasant experience in the past for the sake of learning something from it) and by our attitude to ourselves as suffering at the time of the remembered experience (regardless of whether we were willing to retain unpleasant feelings in our consciousness or, rather, gave in at once to the desire to get rid of them, by means of repression, by refusing to think about them, by diversion of attention to something more pleasant). But more of this under the topic of "vain imaginations."

Before closing this section, I want to compare the foregoing presentation of the part played by imagination in the reconstruction of past experience (being a combination of imaging-as-if-perceiving with empathic imagining) with this role of imagination as it is considered by philosophical analysis in logical studies on the trustworthiness of memory. First of all, in these studies, there is no differentiation of imagination into distinguishably different operations. Second, this total, undifferentiated "imagination" is regarded as productive of fictions, as when we speak of an "imaginary animal." Therefore, imagination is something to be eliminated as far as possible from propositions that purport to be statements of fact about what is remembered. On this account of imagination, the part that it takes in reconstructing the past from memory must be regarded as "delusive" and "falsifying," as it is in the following account by Professor Von Leyden:

> For instance, I remember now how as a boy I was carried home from school with an acute appendicitis. The picture of the scene I remember differs from that which I must have perceived at the time in that I now see myself, in my mind's eye, sitting in the class-room, writhing with pain, and being carried out in my form-master's arms—all aspects of the situation which I could not have experienced in the past on account of my absorption in the agony I then felt. In fact, part of my recollection now would seem to be rather like what my class-mates must have witnessed on that occasion. It is characterized, in the first place, by a feeling of detachment from myself as suffering, and secondly by a number of vistas (e.g., the back-view of my form-master, running down the stairs with me in his arms, though because of his bulky figure I myself am not fully visible) and indeed by an all-round range of visualizing the situation in question; also by a peculiar flavour caused by the projection of the scene against my entire childhood background and by various other contextual conditions and relationships, to say nothing of the extent to which all this may have become permeated by other people's subsequent reports and by experiences of my own on other similar occasions. Of all these novel factors that emerge, the feeling of detachment from my own past self brings out best the nature of the shift of viewpoint which as a rule intervenes between a past experience and any subsequent recollection of it. (9)

This example shows clearly the operation of both imaging-as-if-perceiving and empathic imagining—"all-around range of visualizing the situation," "part of my recollection would seem to be rather like what my class-mates must have witnessed on that occasion" and also seeing "myself, in my mind's eye, sitting in the class-room, writhing with pain, and being carried out in my form-master's arms." Is it really true that these are "all aspects of the situation which I could not have experienced in the past on account of my absorption in the agony I then felt"? And even if it were true, would memory be regarded as more trustworthy if it brought back to mind nothing whatever but an exact replica of the all-absorbing pain? Would the isolated pain be remembered at all, without the filling in of its context and the "projection of the scene against my entire childhood background"? It seems to me that whatever part was played by imagination in the reconstruction of this scene has made more sense out of it than the writer could otherwise have achieved, especially if only the pain is regarded as the trustworthy part of the memory. This is not to say that imagination cannot be both delusive and falsifying in some of its operations, as we see presently, but only to claim that it is not necessarily and by nature so—but then, we are stopped again because the different operations of imagination in these philosophical analyses have not been distinguished, and therefore, its general nature must be regarded as fiction-producing and its part in remembering as delusive and falsifying.

IV

When it comes to understanding the interior lives of fictitious persons or of characters in literature generally, we can no longer avoid considering the operation of *expressive imagining,* for here we are entirely dependent on a writer's talents and purposes and energies. If we confine ourselves to the use of words, leaving other kinds of expressive imagining for a later consideration, we may say that expressive imagining consists of the writer's or speaker's *experimenting* with words, inventing stylistic devices, and *discovering* new interverbal functionings (including leaving some things unsaid), until he has found the clearest and most adequate verbal embodiment of the "what" that he wishes to communicate. As previously noted, it is the felt inadequacy of preliminary embodiments that leads to a reexamination of the "what" by means of empathic imagining, and a reembodiment in words, and so on.

The final embodiment or enactment in words that the writer settles for has also a retroactive effect on the "what" of the communication, "freezing" it in that form. And although the writer can never convince himself that he could not have said it better, he learns to be grateful, in the long run, for the discreteness and numerical limitation of words, for otherwise he could never bring the reciprocal process to a conclusion. The writer, at first struggling with the resistance of language to his purposes, really needs this resistance to help him give a final, definite form to his work. Imagine what it would be like trying to write with, instead of words, an infinite continuum of syllables capable of expressing anything—one could never finish the thing off!

What always amazes us is that in the hands of the great novelist language, with all its limitation, is able to make us feel that we understand the interior lives of invented characters better than we understand the interior lives of some people we see every day—people we can observe, touch, talk with, and do things with; people of whom we can ask opinions; people whose feelings we can ask about directly. The reason of course is that we simply do not bother to thematize and convert into verbal descriptions the many, many clues that most of the people we see every day send us (or that we could elicit from them); we take this trouble only in relation to persons who are important to us. The novelist thus makes up for our laziness and inattention in this respect, when presenting his invented person. But there is another reason. Even in relation to persons who are important to us, there is a limit to how much we can quiz them or trick them into revealing or betraying their interior lives—they may be too inarticulate to tell us, or they may be simply unwilling to tell us. Beyond a certain point, all such attempts to understand real people constitute an invasion of privacy and are permissible, if at all, only in the confidential relationship with a doctor or clergyman or counselor. Here is where the novelist is at an advantage. His invented character may have a certain privacy in relation to the other characters in the story, but not to the novelist himself, and hence not to the reader. Thus we see that the novelist's invasion of the privacy of the fictitious person, and his description to the reader of what he learns there, is a device for getting around the inaccessibility of so much of the data of feeling, and insofar as he is able to make some order out of this "chaotic manifold," he brings to the reader some cognition of feeling that the reader had not had before and could not have gained in any other way. Needless to say, this is true only of the "great" novelists (the "small" ones repeat clichés); the great fictitious characters of literature become the "indirect way" that we must employ, according to Baensch, in approaching the cognition of feeling, including emotion. How far the reader then decides to use this new knowledge for a better understanding of himself and his fellow man, or whether he uses it merely for entertainment (not such a sin as Collingwood would have us believe) is beyond the scope of this discussion.

Expressive imagining is not confined to the description of feelings, emotions, and other conscious states, or of literary characters. It operates just as necessarily, though less conspicuously, in the description of scenes and events (as in giving evidence at an accident), in the presentation of information (as in taking an examination), in the summarizing of somebody else's thoughts (as in writing a term paper), or simply in clarifying one's own thoughts. In all these cases one must use the trial-and-error method of hunting for just the right combination of words; always this involves an element of imagining and also of disappointment, and it may be lightning quick or ponderously slow, depending on the circumstances. Here we run again into the problem of the relation of language to reality. For an essay on epistemology that tries to be as empirical as possible in all areas, it must seem regrettably improbable that a satisfactory answer to this question could ever be found in the logical analysis of sentences, in the presentation of paradigm cases for establishing usages, and in some general theory of semantics. The trouble is that the philosophers of language have abstracted from the total communicative situation as it actually exists only that part that can be treated in terms of logical relations; they have left out precisely the psychological and behavioral context, in terms of which the two communicators understand each other very well. As Joseph Church puts it,

> It is in trying to decipher the semantics of utterances detached from their behavioral contexts that students of meaning have gone astray. Instead of asking what a statement, considered as an objective entity, means, we might better ask what this individual means (or intends) when he says thus-and-so, *and* what this statement uttered by so-and-so means to this listener. It is obvious that we have removed meaning from the level of the word to that of the utterance. Words do not have meanings, but functions. The "meanings" assigned to words by dictionaries are abstractions drawn from the ways words function in various contexts . . . . We must draw a distinction, however, between conventional *signs,* which have a single fixed meaning regardless of context, and *symbols,* which convey information and vary in meaning according to context. In some cases, of course, an utterance can be both a sign and symbol: a forefinger against the lips has the sign-value "Quiet!" but it also can convey that the signaler wants to surprise somebody, that the person to whom it is addressed was about to say something dangerous, that there is something important to be heard that would be drowned out by speech, and so forth. We are saying here that what has long been treated as a logical problem is in fact a psychological one, that the problem of meaning is the problem of how meaningful utterances come to be uttered by a speaker and comprehended by a listener.(10)

The relation between language and reality is treated more carefully in Chapter Five. Here we can only point to the part played by expressive imagining in the long and laborious process of turning a nonverbal reality into a verbal reality, which extends all the way from the egocentric monologue of the child, with its accompanying mobilization toward action in a non-verbal schema, to the dialogue between two adults, which may or may not result in action and which is often engaged in for its own sake, for the sheer joy of communication between people. Egocentrism in the child, says Professor Church, means not selfishness, but being so embedded in one's own point of view, with its partial schematization as a *world,* that one cannot imagine that any other world could exist for anyone else, and therefore "my" world is *the* world, obvious and unquestionable.

Egocentric speech is characterized, as Piaget says, by failure to take account of the listener's orientation and information and to make sure that he is given all the information necessary to understand that is said. It is almost as though the child assumes that he and his listener are surveying a common landscape from a common vantage point and have identical interests and concerns, so that all he has to do is point to a few key features to make himself understood. It does not seem to occur to the child that the other person might possibly have preoccupations of his own or be ignorant of something that the child takes for granted. It is not that the child is unwilling to make himself clear; he will repeat himself endlessly and even patiently—it is just that he cannot understand where the difficulty in communication lies. (11)

Egocentric speech remains in the adult as the interior monologue, but even here, where no listener is involved, there is a constant differentiation between the public world and the private worlds of persons, so that if an imaged Mr. X is addressed, the monologuist must adjust the expression to what he imagines of X's interior life, interests, knowledge, and so on. All so-called creative writing has its origin in the interior monologue in which one "talks to oneself" or talks for one's own benefit; and in its first "try-out," the monologist himself is the critical audience, approving, disapproving, suggesting changes, ransacking memory, giving up hope for anything better. As a help in the discipline of converting the rambling phrase-image-feeling flow of the stream of consciousness into sentences with a beginning, middle, and end, it is important for the writer to imagine a somewhat vague or generalized reader or listener. Indeed, the more specialized the writing, the more important it is for him to imagine correctly the probable background and equipment of the anonymous, potential reader. Even writing down verbal ruminations is a great clarifier of one's thoughts, for as long as such notions are in solution and just floating around in one's head, one may think them to be very fine, important, and beautiful thoughts, just because they are one's own. Once parceled out into sentences, however, each package a struggle with words, laid end to end to make a composition, fixed, rereadable, criticizable—then it is usually a different story! People who turn out to be creative writers must enjoy this endless struggle with words, otherwise they would give it up; and of many people, those who are called inarticulate, it must be said that they literally do not know what they think, because it was too much work for them to put it into sentences. Expressive imagining in the verbal field, or rather, the degree of it that one possesses, is the real difference between these two kinds of people.

What I have called expressive imagining is usually the specific content of what is meant in discussions about "creative imagination" where no distinguishable functions of imagination have been separated, as has been done in this chapter, and where we wish to say something good about imagination, instead of something bad, such as that it is delusive or productive of fictions and unreality. Thus aestheticians and poets tend to praise creative imagination as being the *sine qua non* of art, calling it "the shaping spirit" or the superior reality; scientists admit that it is helpful in some parts of their work, especially in theorizing; and logicians tend to behave as if it did not exist, or that if it did, imagination should certainly not be allowed to contaminate their work. My purpose has been neither to praise nor blame, but simply to take a closer look at the process, to see what jobs it can do and what operations of the mind it is inevitably involved in.

Before looking in the next and final section at the "negative" aspects of imagination—that is, at the mischief it can cause when used mistakenly or wrongly—I should like to close this "positive" portion of the exposition with a few concrete examples of expressive imagining at work, so that we can see how variously it goes about its job.

We have not been able to anatomize imagination without running several times into the topic of dialogue between adults and the curious questions it impinges upon: the existence of other minds, the language of facial expression and gesture, the meaning of words with and without behavioral context, the "filling in," empathically, by the listener of what the speaker has omitted, the mutual adjusting of frames of reference or "worlds" between speaker and listener, and so on. Therefore I have selected four examples that have dialogue between adults more or less as their subject matter. Since I cannot place them before the reader side by side, I will list them in order without prejudice, alternating the two poetic and the two prose specimens.

The first is the second stanza from Yeats's "Among School Children."

> *I dream of a Ledaean body, bent*
> *Above a sinking fire, a tale that she*
> *Told of a harsh reproof, or trivial event*
> *That changed some childish day to tragedy—*
> *Told, and it seemed that our two natures blent*
> *Into a sphere from youthful sympathy,*
> *Or else, to alter Plato's parable,*
> *Into the yolk and white of the one shell. (12)*

The second is a description by Professor Church of what happens when the dialoguists are having an argument:

> But if I am arguing with somebody about American foreign policy, I am behaving only partly toward my opponent and partly toward the topic of discussion. How, then, does the "topic of discus-

sion" exist for us? It can only exist as schemata, mine and my opponent's. When my opponent speaks, he is verbalizing a portion of the schematic framework within which he orders his life. Or, to speak phenomenologically, the topic of discussion exists concretely in the uprush of feelings, and the corresponding changes of body state, that come when the subject is mentioned. In a matter where a person has no first-hand experience, his schemata are the residue of all the relevant verbalizations that he has met and made. If he has thought out his position carefully beforehand, his utterances flow out easily as habitual speech. Otherwise he is forced to translate his feelings into words. He mobilizes himself to speak and his feelings take vocal shape. I, listening, put on his schema and, finding it uncomfortable, try to find words to express its lack of congruence with my own. In other words, my reaction to a statement as true or false, correct or incorrect, inspired or foolish, is contained in the schematic—and therefore affective—arousal it produces. In short, the nonmaterial realms to which language may direct us lie in the schemata of speaker and listener. When a topic is defined, the disputants are mobilized toward each other in such a way that their knowledge of the topic is ready at hand. As they talk, they live—by participation—each other's thoughts. It is not enough to say that a particular set of response habits comes into play, since the disputants are likely to say things they never said before, or to say old things in new ways and in new combinations. What happens in fact is that one person begins to thematize his attitudes, and the other chimes in with elaborations, corrections, objections, or alternative thematizations. (13)

The third specimen is the first and apparently dedicatory stanza of Wallace Stevens's "Notes Toward A Supreme Fiction":

To Henry Church

And for what, except for you, do I feel love?
Do I press the extremest book of the wisest man
Close to me, hidden in me day and night?
In the uncertain light of single, certain truth,

> *Equal in living changingness to the light*
> *In which I meet you, in which we sit at rest,*
> *For a moment in the central of our being,*
> *The vivid transparence that you bring is peace. (14)*

And finally, an example of the failure of dialogue, from Martin Buber's *Between Man and Man:*

> Many years I have wandered through the land of men, and have not yet reached an end of studying the varieties of the "erotic man." . . . There a lover stamps around and is in love only with his passion. There one is wearing his differentiated feelings like medal-ribbons. There one is enjoying the adventures of his own fascinating effect. There one is gazing enraptured at the spectacle of his own supposed surrender. There one is collecting excitement. There one is displaying his "power." There one is preening himself with borrowed vitality. There one is delighting to exist simultaneously as himself, and as an idol very unlike himself. There one is warming himself at the blaze of what has fallen to his lot. There one is experimenting. And so on and on—all the manifold monologuists with their mirrors, in the apartment of the most intimate dialogue! (15)

The first three examples are concerned with what sometimes happens, though not always, when genuine dialogue between adults is attempted; the fourth describes what happens when dialogue fails to occur even though attempted, and why. I have chosen these four examples for the great variety of expressive *means* that they employ. The stanza by Yeats contains an obvious instance of "vivid poetic imagery," thought by some to be the most important device in the arsenal of the poet, and as we noted, called by Longinus *phantasiai,* a useful tool for the rhetorician. This image is so specific that a psychiatrist might even object to it on the grounds that the engulfing of the yolk by the white of the egg indicates a dangerous situation. We know, however, that for the purpose of this stanza Yeats tried to concretize the moment of participation that characterizes the "youthful sympathy" into a single, unforgettable, almost "shorthand" image.

The second example is a scientific description of what happens when the dialogists are opponents in an argument, although nothing indicates that they might not also be friends. The description is long, objective and thorough, and it tries to achieve accuracy by the accumulation of particulars, and by a certain degree of dramatization: "I and my opponent" are shown *doing* certain things.

The Stevens stanza I have included precisely because it is almost devoid of any "vivid poetic imagery" (except for a hint in the second and third lines); on the contrary, it is full of conceptual terms and abstract ideas, a "sin" in the eyes of some modern poetics.

Finally, the selection from Buber is surely prose verging on the edge of poetry—especially if we allow full play to the imagery of "in the apartment of the most intimate dialogue!"—and it uses almost exclusively the method of dramatization.

The curious thing to note is that, in spite of the variety of means employed for the "how" of these descriptions, there is astonishing agreement on the "what"—namely, the nature of dialogue. All four writers agree that when genuine dialogue is taking place there is some kind of *participation* by the dialogists in each other's "being", or "nature", and that, as the Buber selection shows, when this element of participation is missing, the dialogue simply disappears, or rather, collapses into multiple monologue. The Yeats stanza instances the case where the participation is achieved by way of empathic imagining of the other person's hurt feelings. The scientific description shows that even where the dialogue is an argument, if it is to remain dialogue the disputants must put on each other's schemata, and live each other's thoughts by participation, and verbalize the felt incongruences and suggest "elaborations, corrections, objections, or alternative thematizations." Not, in other words, just talk *at* each other, in the manner of Buber's monologuists. What a revealing light this throws on most of what happens in political debate!

The Stevens stanza, in the first three lines, appears to want to distinguish and contrast the way the dialogist feels toward the topic ("the extremest book of the wisest man") from the way he feels toward his partner. The next five lines describe the nature of the participation. In contrast to the Yeats stanza, where the moment of participation is characterized by sudden, youthful sympathy, in the Stevens stanza there is a cool, restful meeting, "in the central of our being," unruffled by the uncertainty of truth, peaceful in its vivid transparence and living changingness. In other words, *this* participation is characterized by what some people would call an "intellectual" or "cool" emotion. But it *is* characterized, it is neither a "commotion" nor a "chaotic manifold." Finally, the Buber selection makes us positively *see* the posturing monologuists—each one using his listener as a mirror in which to observe and admire himself. Very little participation here—and all of a sudden we know why so little genuine dialogue takes place, in even the most intimate apartment of most cocktail parties! In this way, and by such different means, devised by expressive imagining, we learn something that we perhaps did not know before, at least about the participatory aspect of dialogue.

V

Finally, in this little anatomical essay on the operations of the imagination, we must consider the negative aspects of it, the uses to which it can be put that result in illusion, delusion, falsification, and deceit, in the hindering of knowledge rather than in its enhancement. The distinction that needs to be made here is between *errors* of the imagination—what might simply be called *incorrect* imaging and *inadequate* imagining—and the special quality of *vain* imaginations. Nothing in the foregoing analysis of the positive operations of imagination suggests any guarantee that the operations will be successful in the sense of being free from errors and inadequacies. In fact, we noted that both imaging and imagining require a great deal of work, brainwork, to be exact, so that all factors that are likely to impair brainwork are likely to introduce error: laziness, inattention, indifference, fatigue, disease, the effects of some drugs, the absence of strong motivation, such as curiosity or enthusiasm, to learn about the world—and of course, the hereditary quality of the individual brain.

But it is a different story with the special quality of "vain imaginations." That special quality is conveyed very precisely by the double meaning of the word *vain,* namely, simultaneously conceited and futile, proud and useless, vaunting and self-defeating. These meanings in turn imply some kind of *resistance* to knowledge, and this resistance, unlike the above mentioned error-producing factors likely to impair brainwork, indicates a *desire* of the learner to distort, for an ulterior motive, the reality being learned about.

Two kinds of vain imaginations must be distinguished, the kind that proceed from weakness, and the kind that proceed from strength. The first, or compensatory, vain imaginations begin with the early childhood helplessness in situations characterized by insufficient support, protection, and "mothering" love from the grownups, and too great demands made too soon on the child's unknown and undeveloped capacities. Since this is the kind that commonly issues in grownup neuroses and in some psychoses, it is to the psychoanalysts we must turn for an example, although not for the Freudian explanation.

Karen Horney, of all the workers in this field, specifically pinpointed the mischief-making propensities of the imagination in the process that generates neuroses, so let us use her work as my example. According to Horney, the long period of helplessness through which the human child has to pass creates a situation in which, even under the best circumstances, the child experiences a certain intensity of fear and insecurity that would not be felt by, say, a three-month-old puppy or kitten. In the "best" circumstances, of course, the child is helped through this period by the intelligent love and understanding of the parents; but it is much more likely that the actual circumstances are characterized by not only imperfect love and understanding, but in addition, by the assorted neurotic needs and demands of the parents and other grownups projected on the child and determining the adults' various attitudes toward him: domineering, overprotective, intimidating, irritable, overexacting, overindulgent, erratic, partial to other siblings, hypocritical, indifferent, and so on. "As a result, the child does not develop a feeling of belonging, of 'we', but instead a profound insecurity and vague apprehensiveness, for which I use the term basic anxiety. It is his feeling of being isolated and helpless in a world conceived as potentially hostile." (16)

Whereas under more favorable circumstances a child would gradually develop a real self-confidence—confidence, that is, based on his real self—he now begins to cast about for some way of reducing the basic anxiety by creating what might be called an *ersatz* self-confidence, or, confidence based on an *ersatz* self, to give him the feeling of unity, identity, power, and significance he so badly needs.

Provided his inner conditions do not change (through fortunate life circumstances) so that he can dispense with the needs I have listed, there is only one way in which he can seem to fulfill them, and seem to fulfill them at one stroke: through imagination. Gradually and unconsciously, the imagination sets to work and creates in his mind an *idealized image* of himself. (17)

The child's imagination builds this image out of his own personal ingredients, but always in a glorified form, and he and his energies are thenceforth turned toward molding himself into the idealized self, which becomes to him more real than his real self, "not because it is more appealing, but because it answers all his stringent needs." Horney called this process the "comprehensive neurotic solution" and observed that "it promises not only a riddance from his painful and unbearable feelings (feeling lost, anxious, inferior and divided), but in addition an ultimately mysterious fulfillment of himself and his life." This "comprehensive neurotic solution" then launches the young person out on *the search for glory,* which has as its component drives the *need for perfection, neurotic ambition* (for externally confirmed success) and the *drive toward a vindictive triumph* (as a revenge for humiliations suffered from others). The search for glory, unlike the healthy desire for achievement of normal people, is compulsive, indiscriminate, and insatiable, and when it is frustrated, it brings on severe reactions of panic, despair, depression, and self-hatred.

The idealized self which *in imagination* the neurotic person more and more becomes thus sets up in reality a pride system that the therapist can discover by probing the particular values through which the patient's pride is most easily injured, bringing on excessive reactions of shame or humiliation, plus hatred and contempt for his real self.

> Briefly, when an individual shifts his center of gravity to his idealized self, he not only exalts himself but also is bound to look at his actual self—all that he is at a given time, body, mind, healthy and neurotic—from a wrong perspective. The glorified self becomes not only a *phantom* to be pursued; it also becomes a measuring rod with which to measure his actual being. And this actual being is such an embarrassing sight when viewed from the perspective of a godlike perfection that he cannot but despise it. Moreover, what is dynamically more important, the human being that he actually is keeps interfering—significantly—with his flight to glory, and therefore he is bound to hate it, to hate himself. And since pride and self-hate are

actually one entity, I suggest calling the sum total of the factors involved by a common name: the pride system." (18)

Imagination is required not only to set up the pride system, but to maintain it against all attacks from reality, against the "signals" from the actual situation both inside and outside the person which are trying to break through the curtain of illusion that increasingly surrounds him (a curtain that he himself is busily weaving and thickening from a tissue of such further feats of imagination as lying, dissimulation, pretending, distorted or "edited" empathic imagining of other people's feelings, projecting of one's difficulties on others—well, I need hardly go on). In this way (in terms of the formula for animal knowledge given in Chapter Two), the nervous system gradually becomes nonhomeomorphic with the niche in the nonnervous system in which it is supposed to operate, until, in the case of the severe neuroses, we might say that it has become *inoperative,* that is, "wrecked." Animals, by their lack of imagination, are protected from this outcome. They are not plagued by the specifically human form of fear that Horney calls the *basic anxiety,* and they are not tempted by imagined possibilities, luring them on to try to overcome it. They are not tempted to make a pact with the devil:

> The most pertinent symbol, to my mind, for the neurotic process initiated by the search for glory is the ideational content of the stories of the devil's pact. The devil, or some other personification of evil, tempts a person who is perplexed by spiritual or material trouble with the offer of unlimited powers. But he can obtain these powers only on the condition of selling his soul or going to hell. The temptation can come to anybody, rich or poor in spirit, because it speaks to two powerful desires: the longing for the infinite and the wish for an easy way out. According to religious tradition, the greatest spiritual leaders of mankind, Buddha and Christ, experienced such temptation. But, because they were firmly grounded in themselves, they recognized it as a temptation and could reject it. Moreover, the conditions stipulated in the pact are an appropriate

representation of the price to be paid in the neurotic's development. Speaking in these symbolic terms, the easy way to infinite glory is inevitably also the way to an inner hell of self-contempt and self-torment. By taking this road, the individual is in fact losing his soul—his real self. (19)

Imagination, which makes man human for good or ill, and something other than a clever rat, contains this tragic dimension. Yet its full ramifications come into sight only when we change to a theological way of speaking, remembering that the entire process is initiated, and perpetuated from one generation to the next, by the great scarcity of intelligent, foresightful, other-considering love in the world. The cure, to the degree that cure is possible, is "therapeutic disillusionment" for the purpose of permitting the growth and development of the real self, beginning with the acceptance of this hated self and exploring its possibilities and limitations under the sympathetic guidance of the therapist.

When it comes to the vain imaginations issuing from strength, we are dealing with a phenomenon so universal, and so much the object of envy by those who do not have the requirements, that it would be peculiar to call the phenomenon a sickness on the psychological level. Outside the Biblical view of man, it must be considered the reward of life. It is the endless varieties of narcissistic egomania, the self falling in love with itself on the basis of some *real* advantage over others, whether beauty or talent or power or brains or riches or position in society or ancestry or even moral virtues and spiritual attainments—I cannot enumerate the members of such an inexhaustible category. The advantage is real, not falsely imagined, and sometimes not even having to "prove" itself; but it is desired and loved only because it increases and glorifies the self that happens to possess it. To what degree? Think of Alexander the Great, with all his advantages, weeping with frustration because there were no more worlds to conquer. Surely he was not just compensating for the anxieties of childhood helplessness? Or Napoleon, conquering all Europe for his missing four inches? We have only to read the biographies of the "greats" in history to see the distorting effects of their assorted egomanias on their perceptions of reality. And this comes about because the vain imaginations of egomania prevent the knowledge-enhancing functions of imagination we have just studied from being exercised with any degree of precision or thoroughness. Hence the egomaniac's severely restricted knowledge of the probable past, the real present, the desirable future, other people, their thoughts, feelings, desires, needs, and so on. In the Bible, these traits are included under such categories as "blindness", "stiff-neckedness," and "hardness of heart."*

REFERENCES

1. Longinus, *On the Sublime,* translated, with an Introduction, by G. M. A. Grube, The Liberal Arts Press, New York, 1957, pp. 24 and 26.
2. By permission, from *Webster's Third New International Dictionary,* © 1971 by G. & C. Merriam Co., Springfield, Mass., publishers of the Merriam-Webster Dictionaries. Entries under "feeling" and "emotion."
3. Otto Baensch, "Art and Feeling," *Reflections on Art,* Susanne K. Langer, Editor, The Johns Hopkins University Press, Baltimore, 1958, p. 12ff.
4. Baensch, "Art and Feeling," p. 13.
5. Baensch, "Art and Feeling," p. 15.
6. Baensch, "Art and Feeling," p. 10.
7. R. G. Collingwood, *The Principles of Art,* Oxford University Press, London, 1958, pp. 162–163.
8. Susanne K. Langer, *Feeling and Form,* Charles Scribner's Sons, New York, 1953, p. 27.
9. W. Von Leyden, *Remembering, a Philosophical Problem,* Philosophical Library, New York, 1961, pp. 78–79.

* See Appendix Two, Section II.

10. *Language and the Discovery of Reality* by Joseph Church, p. 127. Copyright © 1961 by Joseph Church. Reprinted by permission of Random House, Inc., New York.
11. Church, *Language and Discovery of Reality,* p. 73.
12. Reprinted with permission of the Macmillan Company, from *Collected Poems* by William Butler Yeats. Copyright 1928 by the Macmillan Company, renewed 1956 by Georgie Yeats. With permission of M. B. Yeats and the Macmillan Company of London and Macmillan Canada.
13. Church, *Language and the Discovery of Reality,* p. 130.
14. Copyright 1942 by Wallace Stevens. Reprinted from *The Collected Poems of Wallace Stevens,* 1st Collected Edition, by permission of Alfred A. Knopf, Inc., New York, 1954, p. 380, and by permission of Faber and Faber, Ltd., from the *Collected Poems.*
15. Martin Buber, *Between Man and Man,* The Macmillan Company, New York, 1955, p. 29. Also with permission from Routledge & Kegan Paul, London.
16. Karen Horney, *Neurosis and Human Growth,* W. W. Norton & Company, New York, 1950, p. 18. Also with permission from Routledge & Kegan Paul, London.
17. Karen Horney, *Neurosis and Human Growth,* p. 22.
18. Karen Horney, *Neurosis and Human Growth,* pp. 110–111.
19. Karen Horney, *Neurosis and Human Growth,* p. 39.

Four Action, Limitation, and Possibility: The Animal Value System

I

Back to the animals, now, to see if we cannot find signs of order in the animal emotions and indications of why, in the human case, the feelings and emotions appear as a "chaotic manifold," barely discriminated by the "crude designations" of psychology. To do this we must take a closer look at the second half of our formula for all animal knowledge—the more active part, which states that the encoding must be "in such a manner that the animal's future performance is either perpetuated or modified by its past performance."

We can pause only a moment to peer at the evolutionary meaning of this claim. Surely, the most fundamental prerequisite of knowledge is a kind of *mobility* of the learner relative to the thing learned about, an ability to approach or withdraw, to perform, and to change the performance. Although in an extended sense the plants can certainly be said to "respond" (to good or bad soil, plenty or scarcity of sunlight, water, insects, poisons, cultivation, etc.), they have to *endure* whatever happens to them; and it would seem that such *passiveness* excludes them from the possibility of knowledge in no matter how primordial a form. And similarly for microscopic animals passively carried about by wind and water.

Not to be dogmatic about it, however, we must simply leave open and unanswered the borderline cases of the various botanical tropisms, the cases of the sensitive plants, and the whole question of whether there is anything in the vegetable kingdom even remotely resembling a nervous system or the directive center of a single-celled, mobile animal. A borderline case is also represented by the multicellular animals living as colonies and permanently or transiently attached to sea-bottoms, such as the corals. It may be that the ancestors of these borderline cases represent the bridge between the two major ways in which the antientropic tendency succeeded in establishing life in the niche of the lifeless—the way of mobility and the way of immobility. In any case, the way of immobility precludes knowledge as it is described here, for the entire formula delineates an action, a job done, a task performed.

The task of the animal is to stay alive. If "task" seems too purposive or teleological a way of speaking about the earliest stages of animal evolution, it should be remembered that it is not necessary to imply conscious planning to describe a machine in terms of the task it is designed to perform. A machine is a lifeimitating contrivance precisely because it is expected to perform a task that must be specified, or else its meaning as a collection of interrelated parts cannot be explained. (This may account for our inveterate tendency to call every explanation of how something is achieved in living organisms a "mechanism.") An automobile is a locomotion-imitating machine. When not kept in running condition, it is a curiosity or a piece of junk; and that is how it would look in a nonpurposive description. It is impossible to make sense out of even the simplest life processes without a purposive description. For example, cell metabolism consists of the intake of food, anabolism, catabolism, removal of wastes, replacement of broken parts—and I need hardly point out the resemblance to an automobile kept in running condition by its prudent owner.

The car owner might well wish for the further step in cell metabolism—growth to the point of cell division! So we are not hinting at any "far-off divine event toward which all creation moves" when we say simply that the task of the animal is to stay alive, as long as the limiting conditions permit.

As soon as a task is postulated we are talking about something that can succeed or fail, so that factors in the situation both external and internal to the animal can be graded along a scale according to how much they contribute to the success or failure of the task. Thus at the very foundation of animal existence—as soon as life appears, and as soon as mobility appears, and the task is to stay alive—at this point we run into our first *normative term,* according to which further changes in the conditions of existence can be graded: their survival value or their survival disvalue. Out of the interaction between two sets of variabilities, which to us, at this distance in time, and with our miniscule understanding of the whole process, can appear only as "random changes"—the random changes in the physical world and the random changes in the gene population—there emerges the world's first value system, the survival value system, not to be ignored or depreciated because it is not also the world's last.

It is instructive, because of the sophisticated and sometimes insoluble problems raised by modern value theory and moral philosophy, to take a closer look at this most primordial of all value systems. It already involves all three of the factors mentioned in the title of this chapter: action, limitation, and possibility. In the case of animals (as against the passivity of plants), action means some kind of mobility or performance with respect to situations that confront the animal. The mobility must be of more than one kind—there must be a *repertoire of action possibilities,* otherwise the action could not favor or disfavor survival. The most primitive repertoire of action possibilities is approach to the beneficial situation and withdrawal from the harmful one. And of course the limitation here is death, without which there again could be no *survival value*. As Teilhard de Chardin has pointed out, there must have been, in the earliest days of the *biosphere*, a veritable explosion of life, because the numerical overwhelming of the available space through reproduction had not yet created its own limitation in the competition for substances and in the devouring of life by life. In the great economy of nature, as the saying goes, every form of life limits other life and is itself limited by other life. It is precisely this inescapable condition of animal existence, which we correctly call *inhuman*, that spells out the unambiguous and anything-but-arbitrary meaning of the survival value system.

Ordinarily we tend to think of the development of the higher species out of the lower ones—evolution—as a change in the direction from simplicity to complexity in the anatomical structure and mutual cooperation of *organs*, that is, parts of the body with specialized functions, every one in its time selected for perpetuation or extinction according to its survival value to the animal. But we tend to forget that these new organs would be useless to the total organism if along with them there did not appear a parallel development of some form of sentience, some signal system indicating to the animal the desirability or undesirability of the kind of behavior that the new organ made possible. Let us call this primordial signal system protosentience, subdivided into two kinds, protoeuphoria and protodysphoria, in order not to read too readily into these observable behavior patterns of all animals our own subjective human feelings of pleasure and pain. The fact is that any theory of the emotions that wished to get to the evolutionary fundamentals underlying the origin and function of emotions in animals would have to abandon at the start the attempt to define and discriminate them in terms of human introspective reports, sticking instead to the behavioral and situational characteristics for their proper categorization. The introspective reports of humans are not useless or necessarily prejudiced, as some of the earlier behaviorists believed; when available, they are most enlightening and can be added to the behavioral and situational criteria. However, they are obviously restricted to a very narrow range of the total phenomena that such a basic theory of the emotions would have to attempt to cover.

The very first repertoire of protosentience accompanying the very first repertoire of action would then seem to be a protoeuphoria, tending to cause approach to the beneficial situation, and a protodysphoria, tending to cause retreat from the harmful situation. So it must have been "in the beginning." But when we look at the present denizens of the animal kingdom we see the end products of a groping-and-weeding-out process of both organs and behavior patterns that has been going on for millions of years. We find, then, that even for our contemporaries among the *simple* animals, let alone the complex ones, the repertoires of action and protosentience are much more complicated than the original positive–negative bipolarity. The task of staying alive and of keeping the species alive turns out to be a many-sided one, involving every animal in confronting and dealing with a number of animal-existential situations that might be regarded as prototypical—enormously different as between different phyla, but constituting in each case a gene-inherited solution to a typical existential problem.

II

The foregoing view of the origin and function of protosentience in the animal kingdom, along with the suggestive analogy of the largely preservative hedonism of the higher animals and small children, has led some psychologists of emotion to explore the possibility that a comprehensive evolutionary theory of the emotions might be built up on the concept of a limited number of *primary emotions,* representing the protosentience accompanying a limited number of prototypical animal activities, the latter constituting adaptive solutions to animal-existential problems. (It must be kept in mind that when the word "adaptive" is used in biology it refers to the survival adaptation of the gene population of a species, not the adaptation of individuals, who, by living or dying, achieve that gene population for the species. Later on, when learning enters the picture more obviously, it becomes apparent that within a species the more intelligent individuals tend to outlive the more stupid ones.) The other emotions are then made up out of mixtures of various combinations of the primary emotions, after the analogy of colors that are made up out of mixtures of the primary colors plus black and white. Such a theory has been proposed most recently and most persuasively by Robert Plutchik. (1) After surveying the present state of emotion theory, he noted that there have already been scattered suggestions and hints in the findings of widely separated experimenters, both with animals and children, of the feasibility of a concept of primary emotions. But there has been no systematic following through of this concept because of the difficulty of agreement on which ones should be regarded as primary, that is, simple and unmixed, and which ones as already mixed or derived. To meet this difficulty, Plutchik proposes some criteria of primary, unmixed emotions. They should:

(1) have relevance to basic biological adaptive processes
(2) be found *in some form* at all evolutionary levels
(3) not depend for their definition on particular neural structures or body parts
(4) not depend for definition on introspections (although these may be used)
(5) be defined primarily in terms of behavioral data (or to use Tolman's phrase, in terms of "response-as-affecting-stimulus").

The last phrase is very important, since it brings out the purposive element in animal life. Animals do not react "blindly" to stimulus—they react to increase, decrease, flee from, overcome, the stimuli. Culling and sorting out examples from many workers in this field, Plutchik then proposes *eight basic behavioral patterns,* which he believes satisfy the foregoing criteria and should therefore be regarded as the prototypes of all emotional behavior. The eight basic behavioral patterns, or "emotion dimensions" are as follows: (1) *incorporation,* the taking-in or accepting into the organism of stimuli from the environment which are regarded as beneficial or pleasurable; (2) *rejection,* getting rid of something harmful by expulsion from the organism; (3) *destruction,* overcoming a barrier to some need or action by destroying the barrier; (4) *protection,* avoiding being destroyed, by means of flight, hiding, retreat, freezing, or comparable means; (5) *reproduction,* some form of pulsatile or orgastic behavior associated with pleasure defined as approach and maintenance-of-contact-behavior; (6) *deprivation,* sadness or grief reaction to loss of enjoyed object; (7) *orientation,* reaction to new situation, before it is evaluated as harmful or beneficial, as in startle and surprise; (8) *exploration,* interested motion around the environment to find out its nature.

Two more factors must be taken into consideration before a model or "structural analog" with color sensations can be elaborated: the bipolar nature of the eight primary behavioral patterns (i.e., their qualitative appearance as pairs of opposites) and, within each one, an intensity range from highest intensity to zero. As Plutchik puts it,

> [These primary emotion dimensions] seem to represent bipolar factors or axes with destruction versus protection, incorporation versus rejection, reproduction versus deprivation, and orientation versus exploration. This becomes even clearer if we think of the names used to describe these emotions when expressed in humans. We think of joy as opposed to sadness, acceptance to disgust, anger to fear, and surprise to expectation. These designations are obviously

tentative because of one very important fact, that is, the dependence on intensity of the names given to emotions. Thus the dimension of rejection would include such intensity levels as boredom, dislike, antipathy, disgust and loathing; the deprivation dimension, pensiveness, melancholy, sadness and grief. (2)

An experiment was conducted with thirty college students to try to judge the intensity value of the emotion terms in our language that are usually given as synonyms in the unabridged dictionary and Roget's *Thesaurus*. The subjects were given lists of words describing emotions and asked to range them in terms of the degree of intensity they represent, using a scale of 1 to 11. Combining this information with the eight primary behavioral patterns, we have the following table. (3)

In the last line of Plutchik's chart, in which I have recorded the general hedonic tone of these primary emotions as taken from human introspective reports, the arrows indicate opposites. I reproduce it here so that the reader can compare his own estimations with it and come to appreciate some of the difficulties of naming emotions, even simple, unmixed emotions. It is obvious from this first tentative classification that not all the primary emotions occur in the same intensity ranges. Thus any of the forms of surprise occur at high intensity if they occur at all, whereas the entire acceptance range is lower.

On the basis of this chart, Plutchik constructed a multidimensional model of the emotions, which can best be visualized by wrapping this table around a cylinder of decreasing diameter.

[This model, then, shows] the eight prototypic dimensions arranged somewhat like the sections of half an orange, with the emotion terms which designate each emotion of maximum intensity at the top. The vertical dimension represents intensity, or level of arousal, and ranges from a maximum state of excitement to a state of deep sleep at the bottom. The shape of the model implies that the

Mean Judged Intensity of Synonyms for Each of the Eight Primary Dimensions

Destruct.	Reprod.	Incorp.	Orient.	Protect.	Depriv.	Reject.	Explor.
Rage (9.90)	Ectasy (10.00)	Admission (4.16)	Astonish. (9.30)	Terror (10.13)	Grief (8.83)	Loathing (9.10)	Anticipation (7.30)
Anger (8.40)	Joy (8.10)	Acceptance (4.0)	Amazement (8.30)	Panic (9.75)	Sorrow (7.53)	Disgust (7.60)	Expectancy (6.76)
Annoyance (5.00)	Happiness (7.10)	Incorp. (3.560	Surprise (7.26)	Fear (7.96)	Dejection (6.26)	Dislike (5.50)	Attentiveness (5.86)
	Pleasure (5.70)			Apprehen. (6.40)	Gloom. (5.50)	Bored (4.70)	Set (3.56)
	Serenity (4.36)			Timidity (4.03)	Pensive. (4.40)	Tiresom. (4.50)	
	Calmness (3.30)						
(Neg.)	(Pos.)	(Pos.)	(???)	(Neg.)	(Neg.)	(Neg.)	(Pos.)

emotions become less distinguishable at lower intensities. If we imagine taking successive cross-sections, we keep duplicating the emotion-circle with progressively milder versions of each of the primaries. (4)

The original top-intensity circle is arranged in analogy with the color wheel, so that adjacent emotions are more similar than those that are far removed, and those which are most opposite in quality are opposite each other in the circle, like the complementary colors. In the table, the first and fifth primaries would be opposite, the second and sixth, the third and seventh, the fourth and eighth.

To form mixtures from these primaries and to find the right names for them from our everyday usage is no simple matter. It is somewhat analogous, Plutchik says, to the problem faced by chemists in getting agreement on the exact constituent elements that made up even the simpler chemical compounds, a process that took half a century of experimentation and debate. Many combinations can be made from this table; and when different degrees of intensity of the components are taken into account, it seems obvious that we will quickly get many more possible combinations than are covered by the emotion words in our language. And even more formidable is the problem of getting agreement, since even the primary words and those indicating the simplest mixtures are so carelessly and inaccurately bandied about in everyday speech. The model will yield, as the simplest compounds, three sets of dyads (two-component mixtures), primary, secondary, and tertiary, according as one mixes adjacent, next-after-adjacent, and next-after-next-after adjacent primaries. The quaternary dyads, or mixtures of opposites, are very hard to find names for, since, if they occur at equal intensities, they tend to present the behavior patterns of *tension, conflict,* and *immobilization,* and if persistent and not resolved, usually result in pathological rather than normal emotions. As an example of normal, everyday emotions I will give only the list of primary dyads (adjacent primaries mixed) obtained by the following method. A group of 34 judges (specially selected standard observers) were asked to examine a long list of emotion terms in our language and to indicate which two or three of the primaries they thought were components of them. The words in the following list appeared most frequently as containing the adjacent-mixture terms: (5)

Anger + joy = pride
Joy + acceptance = love, friendliness
Acceptance + surprise = curiosity
Surprise + fear = alarm, awe
Fear + sorrow = despair, guilt
Sorrow + disgust = misery, remorse, forlornness
Disgust + expectancy = cynicism
Expectancy + anger = agression, revenge, stubbornness

An alternative way of presenting these terms is in a table:

Emotion term	Components
Pride	Anger and joy
Love, friendliness.	Joy and acceptance
Curiosity	Acceptance and surprise
Alarm, awe	Surprise and fear
Despair, guilt	Fear and sorrow
Misery, remorse, forlornness	Sorrow and disgust
Cynicism	Disgust and expectancy
Agression, revenge, stubbornness	Expectancy and anger

This is as much of Plutchik's "model" as I have room to present. Although only in its beginnings, it promises to bring some order out of the chaos of both the general talk and the empirical studies about emotions; like the theory of knowledge here being worked out, it is evolutionary in its standpoint.

III

The foregoing excursus into theory of emotions was necessary to prevent our further discussion of the part played by emotions in evaluation and action from getting lost in the mists of vagueness, even though the theory presented will no doubt undergo much improvement in the future. At least we can think of the eight primary emotions as an elaboration of the original repertoire of protosentience, protoeuphoria and protodysphoria, beyond the original approach–withdrawal alternative. This more elaborate emotion repertoire, however, can still be divided into emotions that are by and large pleasant and those which are unpleasant so that, from the standpoint of the animal value system, both the positive and the negative emotions function preservatively (i.e., they promote survival). Animals, it seems, are not hedonists in the same way that grown men and women are: they do not *abstract* the pleasure or pain from the evoking situation in order then to pursue pleasure *per se* or avoid pain *per se.* They pursue or avoid the *situation* per se (or rather, they deal with it by the appropriate behavior, as noted earlier). To animals, the pleasantness and unpleasantness are merely *attention-getting signals,* a sort of stop-and-go system warning them of dangers and attracting them to benefits. Of course the system is not infallible—after all, we human beings utilize it to catch or kill animals by means of all kinds of baits and lures. We present them with a situation that all their heredity and individual experience has led them to regard as pleasant-therefore-beneficial, and then hand them a big surprise. If any animal were capable of doing what human beings constantly try to do, "get rid of" unpleasant emotions by means of repression, diversion of attention, escape into fantasy, blaming others, taking drugs, and so on the result would be disastrous. The famous hedonic calculus would never work for animals, since for them the hedonic tone is not an end in itself but the signal for dealing with a prototypical animal-existential situation.

The problem of the part played by heredity and the part played by learning in the behavior of the individual animal can now be stated somewhat differently from the descriptions of the days of "instincts" and "drives." That which each individual animal inherits, and does not learn, is the specific form which the repertoire of behavior patterns dealing with prototypical situations has developed up to that point in the evolution of his class. But within each species-specific repertoire of action possibilities, the individual animal can behave "intelligently" or "stupidly" according to (1) how quickly and accurately he recognizes the prototypical situation, or any combination of several, that at the moment confronts him; (2) how quickly and accurately he utilizes the signal system of pleasant/unpleasant, or rather, beneficial/harmful, to mobilize appropriate behavior; and (3) to what extent he is able to modify his behavior when (1) and (2) together yield unexpected results. Anticipating now, we note that all three steps involve what we describe in the next chapter as *categorizing behavior.* Step 1 is sensory-perceptual or Gestalt categorizing, step 2 is affective categorizing, and step 3 is the act of reclassification when an unexpected result calls attention to an error in the preliminary classification.

Or, as we put it earlier, in the more cybernetical terms of our general formula, the animal, by *acting* upon his environment, receives a set of "feedback signals" which report to him the success or unsuccess of his performance, causing him to modify his action and thus to receive a modified set of feedback signals, and so on, over and over again. In this manner, providing the animal survives his mistakes, his task of staying alive is facilitated. In these terms, the original subdivision of protosentience into protoeuphoria and protodysphoria may be regarded as the original feedback signals that help the animal to form an *increasingly accurate mapping* of his niche in the nonnervous system (the part of the environment from which he actually receives sensory signals) in terms of *benefit-expectancies* and *harm-expectancies.*

The foregoing picture of the interaction between a nervous system and the nonnervous system in which it is situated helps to explain why "being a classifying machine in a classifiable world" (see next chapter) turns out to have such survival advantage that a general increase in intelligence must almost be expected as we move upward on the phylogenetic tree. The fact that the *individual* animal's "more accurate mapping" is *not* passed on to the next generation does not matter here. What matters is that it helped him to survive to the point of reproducing his kind, including his "better classifying machine"!

That an animal learns from individual experience is now so generally recognized that we are compelled to ask about the further profitability of talking about "instinct" and "instinctual behavior" (as against talking simply about unlearned behavior and learned behavior), recognizing the difficulty of distinguishing them, since on the basis of the unlearned inherited pattern, the learning starts at the instant of birth. Of course the survival value of the learning process depends also on the flexibility and the variety of possibilities in the inherited action repertoire, since the problem is often what to do in the face of a new situation or a variation on the expected one. For example, the most intelligent fish cannot do anything about a too-great rise in the temperature or pollution of its water. Magda Arnold, in her massive study of emotions in men and animals, *Emotions and Personality,* uses the term "intuitive appraisal" for the emotional component of animal behavior and defines an emotion as "the felt tendency toward anything intuitively appraised as good (beneficial) or away from anything intuitively appraised as bad (harmful). This attraction or aversion is accompanied by a pattern of physiological changes organized toward approach or withdrawal. The patterns differ for different emotions." (6)

Now "intuitive appraisal" is certainly an improvement over the operation of "instinct" in suggesting the cognitive element in the emotional behavior of animals, yet it seems to suggest something quite mysteriously different from what goes on in grownup human thinking, whereas in our analysis we will see that it is the same sort of categorizing activity that is later extended to verbal concept formation, abstraction, generalization, and so on. Perhaps we could remove some of the opaqueness from the words "intuition" and "intuitive" by letting them stand for the sort of nonverbal categorizing activity we have just described, whether it occurs in animals, children, or grownups. Such nonverbal categorizing activity makes it easier to understand the process by which "wild" animals can become "tame," by which most domestic animals can learn the ground rules of the house or farm, and by which many, though not all, items of the purely sensory field of perception very early in life acquire an "emotional charge," as Collingwood called it, which may or may not have to be corrected (reclassified) later.

One more thing we can notice from Plutchick's emotion theory before we pass on to the human condition: the relationship between the intensity ranges of the primary emotions and their preservative function for the animal. This function is a combination of attention-getting with perceptual and affective categorizing for the purpose of mobilizing the appropriate action. Therefore, the emotions at very low intensity are not likely to be much help, since they are difficult to discriminate from one another and hard to pay attention to, so that they encourage somnolence, lethargy, and finally sleep, rather than action. At the highest intensity, on the other hand, the emotions tend to reduce perception in general, because of the high degree of physiological involvement (vascular, glandular, muscular, etc.). We acknowledge this circumstance in such expressions as "blind rage" and "blind panic" and the meaning of "being beside oneself" for both ecstasy and grief. Even animals panic on occasion, and when they do they cease to behave preservatively. At least some species of animals, moreover, behave suicidally in certain stages of their reproductive behavior pattern. The conclusion seems to be that the primary emotions function most preservatively for the individual animal in the medium ranges of intensity.

IV

Turning now to the conditions of human existence to see how the emotions function there, we find that, like the entire human race in every generation, we must pass through the needle's eye of the very peculiar condition of human childhood existence. Although many other mammalian children spend a certain period at the start of life under the nurture and care of parents, the sheer length of the human child's dependence, and the severity of his helplessness, makes one wonder how *Homo sapiens* survived the evolutionary obstacle course at all. And yet this condition is only the latest and sharpest aggravation of what has already been observed as a "surprising" evolutionary trend: According to D. O. Hebb, "As we go up the phylogenetic scale, then, we find in mature animals an increasing ability to learn complex relationships, but also, surprisingly, a slower and slower rate of learning in infancy." (7) Hebb is one of the newer breed of behaviorists, as the subtitle of his book indicates: "Stimulus and response—and what occurs in the brain in the interval between them." On the theory of neuropsychological functioning developed in Hebb's book, it is not difficult to understand this "surprising" phenomenon of slow early learning being connected with a large brain. Speaking of brains of the same total cortex size, Hebb said,

[It is to be expected that] the length of the primary learning period will be roughly proportional to the ratio

$$\frac{\text{total association cortex}}{\text{total sensory cortex}}$$

which can be called the A/S ratio. The sensory projection areas are directly under environmental control; and if they are large, with respect to the association areas, and so project a large number of fibers into the association cortex, their control should be quickly established. If the sensory projection is small, association cortex large, the control will take longer; the period of "primary learning," that is, will be long.

But we may have to consider a further factor. With the same A/S ratio, but with different absolute size of brain and a larger absolute number of transmission units in the association areas, it seems that the larger brain might also make for slower first learning. A larger number of transmission units means a greater variability in the spontaneous activity of the association areas, as well as a larger number of synaptic junctions over which sensory control must be extended. It will be recalled that the synapses in question are not the minimum number through which an afferent excitation might conceivably reach the motor cortex. We are dealing with circuits and complex closed systems. An absolutely greater number of transmission units in the association areas would probably make for greater variability in early spontaneous activity and a greater ultimate complexity of organization. Both imply that a sensory control might be established more slowly in a larger brain, even with the A/S ratio constant . . . . The possession of large association areas is an explanation both of the astonishing inefficiency of man's first learning, as far as immediate results are concerned, and his equally astonishing efficiency at maturity. (8)

I cite these neurological details only to discourage any science-fiction daydreams to the effect that man might some day improve the human condition by breeding to increase the size of his brain indefinitely. It is also good to be reminded that the difference between men and animals has first of all an anatomical, rather than a philosophical, basis, and that no amount of metaphysical or theological upgrading or downgrading of the human being will change this fact.

The human infant, then, as compared with a puppy or kitten, enters the world like a handicapped child. He inherits the eight primary emotion dimensions like all other animals, but he remains for a long time as if maimed or crippled when it comes to performing the prototypical behavior repertoire that would make the emotions operate preservatively for him. When he is frightened he cannot run away; when he is angry he cannot destroy the anger-evoking object; he cannot in general approach the beneficial or withdraw from the harmful situation as "intuitively appraised." Indeed, even "intuitive appraisal" is very slow and full of mistakes, needing constant correction by the parents. For a long time the human child neither reads the "feedback signals" correctly nor modifies his performance accordingly, except in the simplest cases, such as moving limbs back from pain, or accepting food. One does not see a six-month-old kitten or puppy bumping into things, pulling things down on itself, falling down, or swallowing dangerous objects, but an eighteen-month-old child will cheerfully do such things repeatedly.

At this point it would be tempting to speculate on the relation between the origins of the earliest, primal human societies and the long helplessness of the human child. We cannot go into that here, but it seems reasonable to surmise, in an *ex post facto* manner, that since a certain number of children with longer and longer childhood helplessness *did* survive, the possession of the larger brain *by the parents* must have been accompanied by an increased exercise of at least two of those functions of the imagination we studied in Chapter Three. Projective imagining and empathic imagining would have enabled the parents to provide foresightfully for the child's future as well as its immediate needs, to avoid the foreseeable dangers on its behalf, and to understand, empathically, the child's feelings. It is likely that some of the earliest societies were matriarchal, the mothers encouraging adult males to assume the roles of hunter and warrior (i.e., food-getter and protector) for the sake of the youngest children; not until it became generally possible to accumulate possessions could the males afford the patriarchal luxury of owning women and children. (9). At any rate, it seems likely that it was the long helplessness of the human child that first introduced the social, other-regarding dimension into the grownup human emotion repertoire, since playing the parental role successfully now demanded long-range, settled attitudes of love, helpfulness, protectiveness, tenderness, and worry, for the sake of another human being, rather than the immediate response to pleasures or pains. This phenomenon can be observed even today, when we see the "natural narcissism" of young people turn into other-regarding love with the assumption of parenthood.

In the first appearance of human conditions of existence, then, as against animal conditions (a matter of hundreds of thousands of years), the whole animal value system of survival, and the repertoire of inherited behavior patterns representing the eight primary emotions with their largely preservative functions, are thrown out of kilter, as it were, by the peculiar conditions of human childhood existence. From the standpoint of the child, the emotion repertoire becomes polarized around the opposition of love and hate, or, more accurately, the love and protection of himself by the parents or grownups as against the hatred and neglect of himself by them, and these two relationships become his chief assurance of survival and his chief threat of annihilation, respectively. From the standpoint of the parents, this polarization around the axis of love/hate becomes the *leverage* they use to teach the child the habits of emotional control and expression sanctioned by their society, that is, the society's value system. Thus they achieve on behalf of the child something more than mere survival, namely, integration into their culture and success as a grownup according to the more generalized and impersonal criterion of approval/disapproval by that society. When the human child has reached the age of such physical independence that he might conceivably survive on his own as a large-brained animal (4–5 years?), he has been introduced by the very agency that had kept him alive to that point—the societal factor—into a world of grownup's values and disvalues about which the ordinary animal knows nothing and to a whole range of possibilities and limitations that the ordinary animal need never face.

REFERENCES

1. Robert Plutchik, *The Emotions: Facts, Theories, and a New Model,* © Random House, New York, 1962.
2. Plutchik, *The Emotions,* p. 109.
3. Plutchik, *The Emotions,* p. 114.
4. Plutchik, *The Emotions,* p. 110.
5. Plutchik, *The Emotions,* pp. 117–118.
6. Magda Arnold, quoted by Plutchik, *The Emotions,* p. 176.
7. D. O. Hebb, *Organization of Behavior,* Science Editions, John Wiley & Sons, Inc., New York, 1961, p. 113.
8. D. O. Hebb, *Organization of Behavior,* pp. 126–127.
9. Ian D. Suttie, *The Origins of Love and Hate,* The Julian Press, New York, 1935, Chapter VIII.

Five The Problem of Correspondence: Devices for Mapping or Picturing Parts of Reality

I

What is knowledge? In answer to this simple question there exists a simple descriptive formula that recommends itself immediately to common sense, to legal practice, and to the easier portions of science; namely, that knowledge is the *correspondence* of our thought with reality. Certainly the negative of this formula—that there exists, in some particular case, a noncorrespondence between the thought and the section of reality involved—describes the existence of error or ignorance rather than knowledge. In guiding our everyday actions we assume that, with some caution, we can obtain information that "corresponds" to the facts. In a court of law, a verdict is overthrown if the story initially given to the jury turns out, on further evidence, not to have "corresponded" to the events as they occurred. In science, the Ptolemaic theory was overthrown when it turned out not to "correspond" to the picture of the planetary system as inferred from astronomical evidence.

But from the positive side we must ask, exactly what does this correspondence consist of, and after what picture or model is it most profitable to think about it? The dictionary tells us that correspondence is "the adaptation, relation, or agreement of things to each other or of one thing to another." It is "relations of congruity, analogy, or similarity." In mathematics, it is the "definite association of certain M members of one aggregate with each member of a second, and of certain N members of the second with each member of the first, called M-to-N correspondence. If M and N are both one, it is a one-to-one correspondence."

There are, then, various degrees and kinds of correspondence, and the history of our subject matter, the theory of knowledge, shows that it is in regard to the degrees or kinds of correspondence between what is still called thought and reality that the different schools make their claims. Radical scepticism essentially claims that there is no way of demonstrating that there exists any degree of correspondence between thought and reality. Therefore we are not justified in calling our putative knowledge the knowledge *of* anything: it might just as well be a useful, systematic delusion. Occupying the middle position are the Kantians and neo-Kantians on the idealist side, and the empiricists and positivists on the realist side.

The Kantians claim there is *some* correspondence, but only what is left over after the raw materials of reality have been fed through and processed by the built-in categories of the human way of thinking, somewhat as meat that is fed through a meat grinder bears the shape of the cutting tools through which it has passed. The unprocessed or raw reality, the thing-in-itself, cannot be known, and therefore anything like a point-for-point correspondence is out of the question. The empiricists agree with the Kantians that *some* correspondence is possible, but they will have nothing to do with that will'o'the wisp, the thing-in-itself, and they tend to regard the categories as characteristics of reality, not of the human mind. Insofar as we discover, and use, the categories correctly, we have already achieved a degree of correspondence between our thought and that reality.

Empiricists, and especially those like the positivists whose model is science, tend to regard the whole correspondence problem as an *approximation process,* a process whereby we increasingly bring our thoughts into better correspondence with reality by weeding out bad ideas and substituting better ones, using experimentation and the verification of predictions as the criteria by which to judge the ideas to be good or bad. Ideas are inventions or constructs, rather like clothing that the tailor cuts and tucks and shapes to fit the individual customer. Although empiricists are modestly aware of the as-yet poor fit in enormous areas of our would-be knowledge, they do not see why there should not be a one-to-one correspondence between our thought and at least the simpler areas of reality. Many of them, indeed, secretly or publicly hope that this "skin-tight" fit will be ultimately attainable everywhere.

At the other extreme are the idealists, both of the subjectivist and objectivist bent, who argue that there could never be even a question of bringing about a correspondence between thought and reality if our minds were not already "tailor-made" to fit that reality. For them even the word "correspondence" is misleading—it is not a question of bringing about a one-to-one correspondence between something and something, but of declaring the *identity* of thought and reality or, as the Hegelians put it, of declaring that the real is the rational, and the rational is the real. Absolute reason is omniscience; and the process of bringing it about is not one of increasing correspondence, but of removing contradictions and exhibiting the increasing *coherence* of all the different parts of thought, as the inherent rationality already implicit in reality is gradually brought to light and made explicit.

It should now be obvious that we are committed to some form of the correspondence theory, since we have already shown that, thanks to the evolutionary weeding-out process, there is always a homeomorphous relation between a nervous system and the niche in reality in which it is able to perform its survival tasks. That lets out both radical scepticism and omniscience as viable options.* And this in turn means that in terms of traditional philosophy, we shall be in competition only with the various kinds of neo-Kantianism and scientific empiricism that occupy the middle position. But when we remember that nothing was known of evolution at the time these "positions" were first taken, the meaning of agreement or disagreement grows nebulous. For example, before genetics and before it was established that learned characteristics are not inherited, what did it mean to talk about "innate ideas" or "a priori knowledge"? We may agree with Kant that the idea-making *process* is "innate" in the sense of being "ready at birth." However, specific ideas, even those as general as space, time, number, or causality, are not innate. These are cultural products, transmitted by language, and they differ quite widely in different times and places. But pleasant and unpleasant—nobody has to learn that at birth! So, was Kant right or wrong?

Since the challenge is to make order out of the chaos of the *matrix world* (the human type of consciousness), we need a model capable of great variation, many levels of abstractness and concreteness, and much correction and expansion. Yet our model must remain a solid model of something, without dissolving either into the misty "nothing" of scepticism or the delusive "all" of omniscience. In this chapter we explore, as a model of the correspondence between what is traditionally called "thought" and "reality" (but which we know is an interaction between one part of reality and another part of reality), that activity by means of which we try to make order out of the chaos of the geographical world: mapping and picturing.

* On the desirability of omniscience, see the final paragraphs of Appendix Two.

II

Erwin Raisz, in *General Cartography,* presents the basic problem of map-making in terms that are peculiarly fitting for both animal and human knowledge: it is the problem of an ant upon a rug, who may have a fairly accurate idea of the design in his immediate vicinity, even though the overall pattern is simply beyond his range of vision.

Let us imagine our ants to have conceived a desire to know the general layout of the rug, to have hit upon the idea of making measurements of the various patches, and to have set some of their number to the task of collecting these measurements and graphically representing the results so that they can be viewed as a whole. Examination of this drawing will reveal a pattern of which they were ignorant before, and no doubt some wise ants will propose various theories to account for the system, nature, and final meaning of this pattern.

How much easier and more simple is the problem of these ants with the rug than ours with the Earth on which we live! A man is a million times smaller in comparison with the Earth's surface than is an ant compared with the largest of rugs, and the richest oriental carpet has a design much less complex than that of the Earth. (1)

Maps and pictures have many points of resemblance and of difference. Thus from the standpoint of categorization, and considering the varying degrees of conventionalization, abstraction, and symbolism used in both maps and pictures, we do well to ask whether a map is a species of picture, or a picture is a species of map. For the purposes of geography, a map is defined as a conventionalized picture of the earth's pattern as viewed from above, with lettering and other symbols added for identification.

Some maps are abstracted and conventionalized to such a degree that the original notion of a picture is hardly recognizable. In many special maps only a single aspect of the picture is retained, as, for instance, in a rainfall map. Moreover, maps commonly exhibit many features which are in themselves not visible, such as political frontiers, lettering, parallels, meridians, etc. (2)

On the other hand, many early maps were scarcely distinguishable from pictures, being only slightly schematized "views" of towns, harbors, islands, or rivers, presented as if looked at from a high place, yet showing quite accurately the spatial relationships of the principal landmarks. It is precisely the advantage of maps that they can use many combinations of pictorial and abstract devices to serve a wide variety of purposes. But accordingly,

> The good cartographer is both a scientist and an artist. He must have a thorough knowledge of his subject and model, the Earth. In representing it he must omit more or less, according to the scale and purpose of his map. This means that he must have the ability to generalize intelligently and to make a right selection of the features to be shown. These are represented by means of lines or colors; and the effective use of lines or colors requires more than knowledge of the subject—it requires artistic judgment. (3)

Highly pictorial maps are not uncommon even today, for instance, those informing the tourist of the historical landmarks of a country or those depicting the location and type of recreational activities in vacation lands.

Characteristic features of all maps are: the scale, the grid, the projection, the content, and the omissions. Whereas ordinary pictures are in a fairly obvious proportional relationship to the object represented (larger than or smaller than life size, miniature, etc.), maps may be in a relationship to the portion of the earth represented that is unrecognizable at sight (say, a million times smaller). Therefore the scale, in inches to the mile or on a graduated line, must be stated on the map. The problem of retaining the resemblance of the pattern on the various reduced scales is solved by means of the grid, a crisscross of regularly intersecting lines, somewhat resembling the "squaring-off" device used in reverse by muralists in order to transfer their paintings from the cartoon to the much larger wall. Only after geographers and astronomers had hit upon the idea that the earth was a globe rotating on its axis could they theorize about what the "natural" grid on such a body would be, that is, one that would best reproduce the pattern of its surface and preserve the proportionality. As we know, they came up with a grid of parallels to the equator intersected regularly by meridians through the poles. Other grids are possible, but not "natural" for a globe. If the earth were a flat disc circumscribed by the river Oceanus, as was thought in Homer's time, the "natural" grid would be regularly spaced concentric circles intersected by radii from its center, similar to our present azimuthal projection. The grid, or coordinate system, ideally supplies an exact location for every item whose two points of reference can be measured.

The desirability of having maps drawn on flat surfaces, because of the cumbersomeness of large globes and the need to present small sections "blown up," give rise to various systems of projection. The three basic ones are: the conic, the cylindrical, and the azimuthal; but numerous variations and combinations exist. The conic projects the natural global grid on two cones set over the northern and southern hemispheres; the cylindrical projects it on a cylinder wrapped around the equator; the azimuthal projects it on two flat planes tangential to the two poles. (Our ants on the rug might have a projection problem the reverse of ours. Supposing them to have guessed correctly that the shape of their rug was a flat rectangle, and supposing that they had only the surfaces of ant eggs to write on, their problem would be to project a large rectangular grid on a small rounded surface.) There are no ideal projections, since they are all by definition systematic distortions of the surface of a globe, and it is only a question of selecting the one that provides the most tolerable distortion for conveying the information for which that particular map is designed.

When it comes to the symbols that are used to convey the content of the map, the ideal of proportionality must be largely abandoned in favor of selective distortion, because many important features of the earth's surface would be microscopic if their true proportional size were used. "On a page-sized map of the U.S., the Mississippi River would be a microscopic, meandering line 1/2000 inch wide. If for this we substitute a wavy line one hundred times wider, we also use a symbol." (4) For mapping purposes, the ideal symbol is one sufficiently resembling the feature it represents that it can be recognized without a legend. But most maps, in addition to this "obvious" symbolism, include a key in the lower right or left-hand corner, to explain the purely conventional symbolism devised for the content of a particular map. When we consider even summarily the types of maps produced, it is easy to be astounded by the amount and variety of information they are able to convey "at a glance," as it were, thanks to the communicative power of their symbolic devices: cadastral maps, showing land ownership; topographic maps, showing relief features; atlas maps, showing large regions; world maps; political maps; city maps; communications maps; scientific maps of various sorts, showing distribution of such characteristics as vegetation, climate, minerals; economic and statistical maps, showing distribution of products, population, races, religions, and so on; navigation maps; oceanographic maps.

Finally, an unobtrusive characteristic of maps often overlooked is the omissions, which are complementary to the selective distortion of important features. What is left out is just as important as what is put in for the intelligibility and usefulness of a particular kind of map, and the temptation to overcrowd a map, especially with many disparate kinds of information, is an invitation to chaos. Since it is impossible to put everything on one map and still have a map, the nonappearance of a given feature on a particular map is no argument for that feature's nonexistence—only a hint that it may have to be sought on another kind of map. Furthermore, some of the symbolic devices used on maps, such as colored areas for different information, are mutually exclusive—they cannot be overlapped.

Maps approach the status of pictures (and vice versa) in the case of aerial photographs, and immediately their special advantages become evident. For although aerial photographs give the exact location of, say, military objectives relative to known features such as rivers and railroads, they are so crowded with irrelevant details and obscured by vegetation, that they are not useful until they have been "read" by experts and converted into more maplike schematic designs. Pictures generally, that is, sketches, diagrams, photographs, and artistic paintings, share to some extent the characteristic features of maps, although these may be used with different emphasis or intent. Pictures all have scale, usually more easily recognized than that of maps, except in the case of microscopic photographs or drawings, or atomic "models," where again the scale has to be indicated, or in the case of astronomical photographs. They all have a frame of reference, but it is less obvious—it is the actual *frame* itself, usually rectangular, which divides the picture into locations: the center, the four corners, the distances from the four sides, and so on. Unless they are the pictures of flat rectangles (like the rug of our ants) they all have a projection which introduces systematic distortion, such as the angle of perspective, the rendering of three-dimensional objects on a flat surface, and conventional shading. Most artistic drawings and paintings have selective distortion (for example, the exaggeration of "important" features and outlining); even the camera, by means of focus and angle and highlighting, can be used in this way. Selective distortion is used in artistic painting to communicate feeling as well as fact, and when it comes to symbolic devices in painting, the possibilities are too numerous to mention. They are in fact all the distinctive particularities by which we recognize the "style" of a special period and of a particular artist. The smooth, serene faces of the Buddha, the elongated faces of medieval saints, the staring, large-eyed faces of the Ravenna mosaics—these are instances of selective distortion that have become almost as conventional as map-making symbols for rendering the content of a particular culture. Although these characteristics do not have a reference as explicit as that indicated in the "key" of a map, they can always be used by artists to evoke whatever in fact they do invoke. And of course all pictures, drawings, posed photographs, and artistic paintings have omissions, the complement to selective distortion, and often the chief secret of their success: for, as with maps, it is impossible to put everything into the same picture and still have a picture instead of chaos.

III

Let us now bring forward from Chapter Two the formula for animal knowledge in its widest application, in order to look at it more closely, see how it must be modified for the human case, and observe how the mapping or picturing model of the correspondence fits into the overall construct we get.

> Whenever there is an interaction between a nervous system and the nonnervous system in which it is situated, such that messages from the nonnervous system (light, sound, heat, pressure, chemical action) are encoded in the nervous system, and in such a manner that the creature's future performance is either perpetuated or modified by its past performance, knowledge of some sort is taking place.

The word "interaction" covers a complicated succession of events, even for the simpler organisms, something more like input (signals), encoding, decoding, output (actions), feedback (signals), further encoding and decoding, modified output, modified feedback, and so on, over and over again, as the organism, gathering clues from the environment on the success or unsuccess of its own past performance, "seeks a new equilibrium with the universe and its future contingencies." (5) Furthermore, the feedback "may be as simple as that of the common reflex, or it may be a higher order feedback, in which past experience is used not only to regulate specific movements, but also whole policies of behavior." (6)

"To encode" is defined by the dictionary as "to translate from the ordinary language into a system of words or other symbols arbitrarily used to represent words," and of course to decode is to translate back to the ordinary language. Obviously, then, when "encoding" and "decoding" serve to describe any part of the operation of the nervous system, they must be interpreted as being used in an extended sense and really ought always to be written in quotation marks. But even Theodosius Dobzhansky, writing "Man's biological heredity, like that of any other organism, is encoded in DNA, carried mainly in the chromosomes of cell nuclei, and transmitted by parents solely to their children and to progeny in the direct line of descent," used the word "encoded" in an extended sense. The geneticist himself indicated his recognition of this when he further said that "Human DNA has merely a different arrangement of the four nucleotides, the four 'letters' of the genetic 'alphabet,' which compose also the DNA of a fish, an insect, or a plant." (7)

In this essay, the terms "encoding," "decoding," "feedback," and "policy-feedback," when used in relation to the nervous system, must not be interpreted as favoring or disfavoring any special theory of neurophysiological activity in the brain. Whether in the long run this activity turns out to be more like the "switchboard" theories or more like the "field" theories, or some much-modified combination of the two, the total task or job performed by the nervous system in its interaction with the nonnervous system would still be describable in these terms from information theory, used in an extended sense. With this understanding, we can drop the quotation marks.

In the most elementary form, then, to encode is to translate information about the perceived sensory discriminations in the environment into neuron activity in the brain, and to decode is to translate this activity back into sensory information about the environment, the brain's "niche" in the nonnervous system. The "key" to this process must be regarded as built into the brain and hereditary, since the conscious percipient is unconscious of any such activity. This unconsciousness is absolutely essential to the success of the entire process, for if the percipient received any messages (e.g., "I am firing now!") from the neurons that transmit the messages received at the sense organs, the former would "scramble" the latter out of all recognition as a pattern or gestalt. It would be as if, in a real telephone exchange, the wires that connect the two parties to a conversation decided to go in for some conversational activity of their own. Thus the unconsciousness of neuron activity in the brain, so often regretted by both philosophers and physiologists as creating an "unbridgeable gap," a permanent hiatus, between the "mental" and the "physical," is really a blessing in disguise—a creature whose visual perception of his environment was constantly interfered with by a parade of scintillations marching up the optic nerve like automobile headlights on a freeway at night could not long endure in the weeding-out process of evolutionary survival.

The perceptive reader will have noticed by now that our formula for animal knowledge bears a certain superficial resemblance to the stimulus–response arc of behaviorist psychology, but this similarity should give him neither alarm nor comfort. For one thing, the simple reflex arc does exist, both in men and in animals. Since, moreover, it can be conditioned and reconditioned, the formula must account for that possibility. For another, it should by now be admitted that the behaviorists themselves have become much more sophisticated than they were in the days of Pavlov's dogs, finding it necessary to hypothesize cognitive processes even in the case of animals to account for some aspects of their learning behavior. The authors of *A Study of Thinking* call this a "revival" of interest in the "Higher Mental Processes," and partially explain the revival thus:

> Partly, it has resulted from a recognition of the complex processes that mediate between the classical "stimuli" and "responses" out of which stimulus–response learning theories hoped to fashion a psychology that would by-pass anything smacking of the "mental." The impeccable peripheralism of such theories could not last long. As "S–R" theories came to be modified to take into account the subtle events that may occur between the input of a physical stimulus and the emission of an observable response, the old image of the "stimulus–response bond" began to dissolve, its place being taken by a mediation model. As Edward Tolman so felicitously put it some years ago, in place of a telephone switchboard connecting stimuli and responses it might be more profitable to think of a map room where stimuli were sorted out and arranged before ever response occurred, and one might do well to have a closer look at these intervening "cognitive maps." (8)

I hardly need point out how "felicitous," for the theory of knowledge being presented here, is this description of what happens between stimulus and response—if indeed there is any overt response, and if indeed the stimulus may not be strictly "mental." For, aside from Tolman's fortunate use of the word "maps," his description shows that even the animals may be liberated from the strict determination of their behavior by the stimulus–response arc and are allowed to exercise, after their own fashion, such "Higher Mental Processes" as concept formation, deliberation and choice, selective attention or "interest," short-range goal seeking, and the adoption of policies of behavior in the future on the basis of the remembrance of things past. As for human grownups, the entire inherited emotion repertoire, as modified by culture, lies between stimulus and response.

IV

If now we wish to denominate more specifically what actually distinguishes human thinking from animal thinking, and human knowledge from animal knowledge, we must return to our consideration of one of the "overdeveloped" activities of the human type of brain—namely, the imagination—and see how this in turn affects, for example, concept formation and concept utilization. Already in Chapter Two we noted, with the authors of the book just quoted from, the seeming paradox in man's sensory situation; that is, man's possession of discriminating capacities which, if fully used, would make him a slave to the particular. We also noted that this seeming paradox was resolved by man's capacity to categorize. "To categorize is to render discriminably different things equivalent, to group the objects and events and people around us into classes, and to respond to them in terms of their class membership rather than their uniqueness." (9) But we said further that this paradox was not resolved for imaginative contemplation, since the discriminable differences did not disappear, but merely allowed themselves to be classified in innumerable ways, producing in man a tension between the hope of order and the threat of chaos. We called this tension his basic epistemic motive. The several operations of imagination, it will be recalled, were then distinguished and described under six headings: imaging-as-if-perceiving, impressionistic imaging, projective imagining, empathic imagining, expressive imagining, and vain imaginations.

To categorize, in the presence of language, we must be able (1) to perceive, or image-as-if-perceiving, or both, a cluster or array of objects or events, in order to "see" their resemblances and nonresemblances; (2) to select and label with a word or phrase, the main features of the resemblances that would justify a class membership; (3) to image-as-if-perceiving, and perhaps also to imagine projectively, further instances of members and nonmembers of the class, or not-yet-existent potential members of the class, in order to explore the range of the category and facilitate future recognition of members and nonmembers, and in this way (4) to "formulate," or "learn" or "acquire" the *concept* for which the word stands. Concept formation, at the very least, divides up perceived reality into sections that are namable, definable, and recognizable. The word or phrase that is used may exist in one's language, or it may have to be invented (with the help of empathic and expressive imagining), but even if it exists, it *was* once invented. Language is then, at the very least, the repository of the accumulated expressive inventions that a human community, in the course of hundreds of generations, has used to divide and subdivide its perceived reality into manageable pieces.

The process of categorization involves, if you will, an act of invention. This hodgepodge of objects is comprised in the category "chairs," that assortment of diverse numbers is all grouped together as "powers of 2," these structures are "houses" but those others are "garages." What is unique about categories of this kind is that once they are mastered they can be used without further learning. We need not learn *de novo* that the stimulus configuration before us is another house. If we have learned the class "house" as a concept, new exemplars can readily be recognized. The category becomes a tool for further use. (10)

Equivalence categorizing presupposes a previous kind of categorizing, namely, *identity* categorizing, which regards each individual member of the class as being and remaining itself throughout its candidacy for the class. "Without belaboring the obvious, identity categorizing may be defined as classing a variety of stimuli as *forms of the same thing.*" (11) This

French Provincial chair, which I see from so many angles and under such different lighting conditions, I continue to regard as "the same chair," and I would probably continue to regard it so, even if it were reupholstered, repainted, or had a broken leg, for purposes of its membership in the classification "chairs." Not so, however, if it were reduced to a heap of ashes, although I might still lingeringly refer to the ashes as "my former chair." Identity categorization may possibly be inborn (hereditary), or else it is learned very early in life. Its development, however, does depend to some degree on learning, for, according to Piaget "At certain stages of development, an object passed behind a screen is not judged by the child to be the same object when it emerges on the other side." (12) Identity categorizing (the same thing) and equivalence categorizing (the same kind of thing) both depend on one's regarding certain selected properties of objects as "criterial and relevant" while others are regarded as irrelevant (selective distortion and omissions). Equivalence categorizing assumes its greatest importance in the problems of generalization and universality, whereas identity categorizing dominates the problems of dealing with change and particularity—for example, we regard as "the same person" a growing creature whom we have to classify at various times as "a child," "a teenager," "a young man," "a middle-aged man," "an elderly person," not to mention the extremes of "a foetus" and "a corpse."

In the absence of language or other signaling devices, whether a person "possesses" a category must be inferred from a common response, behavioral or emotional, to an array of objects or events or situations. "An air-raid siren, a dislodged piton while climbing, a severe dressing down by a superior may all produce a common autonomic response in a man and by this fact we infer that they are all grouped as 'danger situations.' Indeed, the person involved may not be able to verbalize the category." (13) Of the three broad classes of equivalence categories distinguished by Bruner et al., namely, *affec-*

tive, functional, and *formal* categories, the first, affective categories, are prominently characterized by difficulties of verbalization.

> The difficulty appears to lie in the lack of correspondence between affective and linguistic categories. As Schachtel (1947) and McClelland (1951) have suggested, categories bound together by a common affective response frequently go back to early childhood and may resist conscious verbal insight by virtue of having been established before the full development of language. (14)

We are reminded of Otto Baensch's description of the feelings and emotions as a "chaotic manifold" that resists conceptual treatment "beyond the crudest designations," and of Collingwood's description of the preverbalized or preexpressed feelings as "commotions" rather than "emotions." Nevertheless, affective categorization does take place—apparently from the most primary of epistemic motives: that of making order out of chaos. It is difficult to see how a feat of manipulation such as is usually meant by the phrase "control of one's feelings" could be brought about unless there were already some degree of categorization of the specific feeling, some defining attributes by which the onset of the specific feeling could be recognized (e.g., for anger, the provoking situation, the "hot" bodily sensation, and the impulse to strike). Successful manipulation would also require the calling to the rescue, by the person not wishing to succumb to the emotion in question, of strategies previously used for dealing with it, such as reasoning (reappraising the provocation), appealing to higher values, invoking one's ideal self-image, or heeding common-sense warnings against the known consequences in action and feeling of allowing oneself to be "overwhelmed" by that emotion. Self-mastery, which surely distinguishes most adults from both children and animals, is thus a partially cognitive practical enterprise. It involves projective imagining of a future situation, steps to be taken to avoid it, and a certain conceptual grasp of the specific feelings involved, even if these do not go very much "beyond the crudest designations." Similarly, a certain conceptualization of feelings in terms of the provoking situation and the behavioral accompaniments is necessary, if we wish to arouse specific feelings, or avoid arousing them, in other people. As for the communication of feelings from one person to another, or to oneself for clarification, we have already seen how empathic imagining and expressive imagining reciprocally urge each other on for this purpose—how a first attempt at expression, being found inadequate, leads to a more thorough empathic imagining of that part of the feeling-quality we wish to express, followed by an attempt at a better expression of it, and so on. The language, after all, does contain emotion words, such as fear, anger, love, hate, joy, sorrow, surprise, and expectation (to name only the most primary ones); and we learn to use them designatively, just as much as object words and sensation words.

It is probably only to be expected that affective categorizing is easier for the emotions that are primary, unmixed, and relatively intense, simply because they are found in, and describable in terms of, standard evoking situations with standard behavioral accompaniments, including the physiological ones. We can observe such emotions best in animals and in small children who have not learned to inhibit, control, or dissimulate them, and that is undoubtedly why psychologists deal largely in such "crude designations." To pass beyond these to the complex mixtures of socially and culturally conditioned and controlled feelings experienced by adult human beings is a job for the artist in words, using all the devices invented by expressive imagining, including the "emotional charges" on certain images and the emotional connotations of certain concepts themselves. This does not mean that the literary arts are nothing but refined psychology, for to charge this is to ignore precisely the differences in their subject matter, their goal, and their expressive means. The psychologist's goal is *theory,* his means are the simplified situations that generate concepts, and these together determine what subject matter can be treated. The artist's subject matter is *total man,* he who lives in the *matrix world,* the space-time "lebensraum" extended from perception by imagination, in which memory and anticipation play as much a part in feeling-evocation as the present, and in which the "behavioral accompaniments" may be so subtly controlled and dissimulated as to defy observation. As Baensch remarked about the problem of describing the quality of the more subtle feelings in words:

> We are really confronted with a chaotic manifold from which no internal principle of order can be derived. Nothing therefore avails us in life and in scientific thought but to approach them indirectly, correlating them with the describable events, inside or outside ourselves, that contain and thus convey them; in the hope that anyone reminded of such events will thus be led somehow to experience the emotive qualities, too, that we wish to bring to his attention. (15)

The difficulties of verbalization connected with affective categorization have frequently led epistemologists to exclude the emotions from the field of cognition entirely, on the grounds of their being too transient, too arbitrary, too ineffable, and too private. An example of this is the positivists' dichotomy: emotive/cognitive, in their analysis of the meaning and use of language. "Emotive" use of language is supposed to be devoid of cognitive content, whereas "cognitive" use of language is supposed to have to undergo a sterilization of its emotive content before its verifiability can properly be judged. The simpler and stronger emotions, indeed, are to some extent categorizable, namable, and definable in terms of publicly recognizable attributes; but this is only a beginning in the right direction.

The other two kinds of equivalence categorizing, the functional and the formal, are what is generally supposed to occupy and rule the entire field of cognitive activity. They correspond (very roughly) to what we ordinarily call "commonsense" knowledge and "scientific" knowledge. No hard-and-fast line can be drawn, however, since most scientific knowledge began, at least in the early days of science, in a commonsense problem demanding a solution. The criterial attributes of functional categories take the form of concrete and specific task requirements: "'things large enough and strong enough to plug this hole in the dike.' Such forms of defining response almost always have . . . a specific interpolative function ('gap filling') or a specific extrapolative function ('how to take the next step')." (16) The authors of *A Study of Thinking* do not use the concept of imagination at all, either in its undifferentiated pejorative or in its undifferentiated commendatory sense. Nevertheless, it is obvious that such "functions" as "gap filling" and "how to take the next step" involve to some extent both imaging-as-if-perceiving, and projective imagining, as we have analyzed them in Chapter Three, using the example of constructing a thing or driving a car. "Houses" and "garages" and "chairs" are examples of functional equivalence categories, their criterial attributes taking the form of "something to live in," "something to keep your car in," and "something to sit on."

The process of passing from functional categories to formal categories involves an act of projective imagining generally described by the term "abstraction" but actually consisting of *subtracting* from the practical situation those features which can be specified independently of it, and imagining them as the defining attributes of a category or class.

> The emphasis of definitions is placed more and more on the attribute properties of class members and less and less on the "utilitanda properties." . . . The development of formalization is gradual. From "things I can drive this tent stake with" we move to the concept "hammer" and from there to "mechanical force," each step being freer of definition by specific use than the former. (17)

Mathematics is a good example of a system of formal concepts originally developed from practical situations but now pursued as an end in itself, regardless of any "fit" with empirical problems. The chief advantage of formal categories is precisely the generality that is built into them by the process of abstracting *from* specific instances, so that what explains Newton's apple also explains why people do not fall off the earth, why the planets move as they do, and so on. Formal categories can also be constructed concerning things that do not exist (centaurs); concerning things that do not now exist but may possibly exist in the future (female presidents of the United States), and concerning things that ought or ought not to exist (an ideal form of government, warfare among men). The process of categorizing is the same whether the materials being categorized are perceptual or conceptual, affective or imagined; and, of course, many objects or events can be cross-categorized in all four ways: the word "father," for example, is the name of a category whose criterial attributes can be affective, functional, ideal, and formal, all at once.

Beyond mentioning that the world must contain the discriminable features that make it potentially categorizable in many different ways, most of which we do not utilize, the authors of *A Study of Thinking* sidestepped the traditional "metaphysical" questions that have been raised by this fact, although they call their own approach radically "nominalistic" and in agreement with contemporary scientific philosophy. In this essay on epistemology we will hear without panic all the empirical evidence concerning knowledge we can find, and with any luck, perhaps even such a piece of name-calling as "nominalistic!" may turn out to be a hangover from a useless controversy of the past. The critical contraindications of this experimental approach to cognition for such theories as Plato's doctrine of Ideas and Forms and for Aristotle's "categories of Being" would even now seem to be considerable, but it will be more appropriate for us to take them up when we investigate the possibilities of validating ontological or metaphysical thinking at all, in Chapter Seven.

For the present, it is enough to observe, in a philosophically noncommittal way, that from the empirical evidence so far available the brain appears to be a classifying machine operating in a classifiable world, that these two appear to be made for each other, and that there are many different types of classifying machines and many different ways in which the world allows itself to be classified. This observation, however, should hardly surprise us or cause us to exclaim over Divine Providence or Preestablished Harmony, if we always remember to return to the evolutionary outlook that we have all along insisted must be maintained in relation to animal knowledge at its widest application. In the beginning was the classifiable world, and in due time it produced out of its own materials and "in its own image" a little classifying machine; and to be a classifying machine in a classifiable world turned out to be so advantageous to survival that thereafter any gene changes that improved the classifying machine survived and those that worsened it were eliminated. But that is only another way of saying what we have already said: that all the nervous systems that have survived up to the present time have done so because they were sufficiently "homeomorphic" with some niche in the nonnervous system to derive a survival advantage from it.

And what exactly is the survival advantage of being a classifying machine in a classifiable world? Our "nominalistic" authors of *A Study of Thinking* leave us in no doubt about that. Although they are reporting on experiments with human subjects, it should be quite clear by now that the chief difference between animal and human brains is the enormous increase in extent and variety of categorizing behavior of which the human brain is capable, thanks to its sheer size and to that "overdeveloped" capacity, the imagination. The human brain, in other words, is a super-classifying machine, and I close this section with a mere listing of its advantages, as indicated by Brunner et al. (18):

(1) "By categorizing as equivalent discriminably different events, the organism *reduces the complexity of its environment.*"

(2) "Categorizing is the *means by which the objects of the world about us are identified.* The act of identifying some thing or some event is the act of 'placing' it in a class. . . . When an event cannot be thus categorized and identified, we experience terror in the face of the uncanny."

(3) "The establishment of a category based on a set of defining attributes *reduces the necessity of constant learning.* For the abstraction of defining properties makes possible future acts of categorizing without benefit of further learning. . . . Learning by rote that a miscellany of objects all go by the nonsense name of BLIX has no extrapolative value to new members of the class."

(4) The act of categorizing inherently provides *direction for instrumental activity.* "To know by virtue of discriminable defining attributes and without further direct test that a man is 'honest' or that a substance is 'poison' is to know *in advance* about appropriate or inappropriate actions to be taken."

(5) Categorizing permits the *ordering and relating of classes of events.* For we operate with category *systems,* "classes of events that are related to each other in various kinds of superordinate systems. We map and give meaning to our world by relating classes of events rather than by relating individual events. . . . The moment an object is placed in a category, we have opened up a whole vista of possibilities for "going beyond" the category by virtue of the superordinate and causal relationships linking this category to others."

(6) Much of our categorizing is anticipatory and future-oriented; it is a way of mobilizing ourselves toward events and things we have not yet experienced, although they might occur under certain conditions. This is done by means of the "empty category," which we create by combining potential defining attributes. Then we look for, or we create, actual members of this class, or we try to prevent them from occurring—for example, a cure for cancer, or an atomic war.

The last example (my own) is well within the province of projective imagining and serves to remind us of the tragic ambiguity inherent in man's possession of an overdeveloped imagination. Man no doubt outwitted the other animals because his brain is a superclassifying machine, but whether it will help him to survive *with his fellow men* is another question, with which we struggle in the next chapter, under the problem of evaluation.

V

In this section we must return to the problem of the correspondence, in order to see how the mapping or picturing "model" of the correspondence fits in with the description of cognitive activity given in the previous section.

The correspondence we are talking about, it will be recalled, is not between two "somethings" prematurely and falsely dichotomized in common sense and even in philosophical talk as "thought" and "reality," but between two different parts of reality: the one able to receive messages, encode them as knowledge, and act accordingly, and also to send messages; the other able only to be the source of messages, not to receive and encode them. We called these two parts of reality broadly the nervous and the nonnervous systems, respectively. Now the existence of such a large variety of specific nervous systems in the phylogenetic tree, each with its strictly limited set of receiving organs for messages from strictly limited sections of the nonnervous part of reality, would seem to militate against the possibility of a one-to-one correspondence for any of them, even the overdeveloped, embarrassed-by-riches, nervous system of man. We have only to imagine for man a new sense organ, such as one that would inform him by a "ping" or a "flash" that a cosmic ray has passed through his body, or to imagine the extended reception of an existing sense organ, such as being able to "see" radio waves, or to "hear" ultrasonic waves. In that case the classifiable world would certainly "look" quite different to contemporary man, as it also does now to one who is born blind or deaf.

It is of course possible to argue for a one-to-one correspondence *in principle achievable* by man's invention of mechanical sense organs to supplement his biological ones, as he has done in the case of all the devices he has invented to receive the messages from the nonvisual parts of the electromagnetic spectrum. We hardly know what such a "limiting case" would mean, however, since (again in principle) man and his total receiving mechanisms could be imbedded in some all-pervasive element of reality which would make no discriminable differences on any of them. ("The fish will be the last to discover water," quoted by Bruner et al.) I only mention this "limiting" case to show that the impossibility, or, let us say, the extreme unlikelihood, of man's knowledge being in a one-to-one correspondence with the nonnervous parts of reality (i.e., the sources of the messages) does not at all invalidate what knowledge he does have. This knowledge, always occurring in a context of mystery or as a foreground of knowledge against a background of mystery, still makes man accountable. The idea of a one-to-one correspondence is the property of a psychophysical parallelism and a naïve realism about the world that still operates in common sense and in the simpler parts of science. In this realm, every sense "datum" is regarded as corresponding to some nonsensory physical "cause," and it is thought that in this way there is gradually built up in the brain a point-for-point *replica* of the outside world, like the replicas of ancient temples found in museums, only on a suitably miniature scale (to fit inside the head), and in the medium of the "mental," whatever medium that might be.

This leaves us with a relationship consisting of some correspondence and some noncorrespondence. We found that the same relationship also obtained between a map of the earth's surface and the actual surface itself, and between a picture and the subject matter pictured. Neither a map nor a picture is a point-for-point replica of its subject matter—which is precisely what enables each to perform the services it does perform for those who use maps and pictures.

> The good cartographer is both a scientist and an artist. He must have thorough knowledge of his subject and model, the Earth. In representing it he must omit more or less, according to the scale and purpose of his map. This means that he must have the ability to generalize intelligently and to make a right selection of the features to be shown. (19)

Simplification of the earth's surface—leaving out irrelevant details—is certainly one of the chief objectives of map making. And as we have seen, simplification of the environment is one of the chief benefits made possible by man's ability to categorize, to treat different sense patterns as "forms of the same thing," and to treat discriminably different things as if they were equivalent. Our categorizing on the sensory level, then, our arranging the constantly shifting flux of our sense perceptions into a pattern of forms of the same thing (identity categorizing) and the same kinds of thing (equivalence categorizing), can be thought of as a preliminary "mapping" of the *matrix world,* that space-time *lebensraum* in which each of us lives, consisting of sensory perceptions extended by legitimate uses of imagination; and the purpose of this simplification and schematization of the matrix world is the same as that of geographical mapping: to help us to get around in that world, and to get what we want in that world. (Language, and especially the *mother tongue,* can then be regarded as the preliminary verbal mapping of the matrix world, by a child living in a community of interacting people, who pass their communal way of mapping their communal matrix world from generation to generation and thus convert what would be a series of private worlds into a public world. Systematic distortion introduced by language is considered again in Chapters Seven and Eight.)

Now this preliminary mapping of the matrix world, both verbal and nonverbal, is far from being infallible, and it does generate certain mistakes because of the superficial resemblances between things really different and because of the placing in the same class of things that turn out by further experience to be not equivalent. These mistakes are correctible, however, and they are in fact corrected during the process of education, by reclassification, by the learning of new classes, by the learning of superordination and subordination among classes (genus and species), by the learning of external relationships between classes and systems of classes, and so on. The fallibility of the preliminary mapping is thus no cause for accusing the sense perceptions of "deceiving" us, since it is further and more refined application of the classification of discriminable differences that produce the corrections. It is the same with geographical maps. Many early maps of the "world" were quite correct in the vicinity of the Mediterranean, but utterly wrong in the farther reaches of Europe, Asia, and Africa—and rectification was achieved not by denouncing mapping techniques but by applying them more thoroughly in the less known regions.

It might be more suitable then to talk, not about a point-for-point replica of the "objective" world being gradually built up inside the brain out of the mental building blocks of elementary sensations, but about a "model" or "schema" of the entire world of human experiences, subjective as well as objective, being gradually built up inside the brain. In our projected "model," the elements would be categories and their relations would be collected by means of interactions, explorations, inventions, and subsequent mappings or picturings of the same, in many, many different regions of experience and on many, many levels of generality and particularity, abstractness and concreteness. But this "model" or "picture" or "schema," being gradually built up in the brain in the process of acquiring knowledge, far from being a point-for-point replica of the world-minus-man, or the nonnervous part of reality, would necessarily repeat and conform to all the systematic distortions and selective distortions and omissions we noted were inherent in the very process of mapping and picturing, and, in fact, the model's intelligibility and usefulness as a guide to action for the brain would depend on this being the case. A fish's "model" of the fish-world, built up out of fish-mappings of the fish-terrain, would be of no use to a bird. Furthermore, this human "model" would be in a constant though gradual process of change, revision, correction, and expansion, as newer, more accurate mappings from various regions of experience arrived, and as new regions of experience were explored, discovered, even invented (interplanetary travel, e.g., or a new form of art). This change, revision, correction, and expansion of the "model" would occur both in the course of the biography of every individual and in the "biography" of the whole human race, thus producing a cognitive equivalent of the biological principle that ontogeny recapitulates philogeny: every individual, in building up his "model," would be privileged, insofar as he was willing and able, to recapitulate the "model" that represents the cumulative experience of mankind up to that point.

It is perhaps an unnecessary quibbling to make this distinction between a model and a map, for is not a model a kind of three-dimensional map, and is not a topographical map of the world a kind of miniature model of the earth? The trouble is that the model that is gradually built up in the brain is composed of so many different kinds of mappings with so many different symbolic devices and contents, that it is hard to see it as just another map—it would in any case be more like a topological than a topographical map. And especially that part of the model we call "the common-sense world" has about it a certain three-dimensionality and picturability that we associate more with models than with maps. In fact, the word "picture" would do for the model in the brain, except that it seems to militate against the other senses, against non-sensory relations, and against the depth and time dimensions. It has in its favor, however, its venerable association with the term "world view," or *Weltanschauung*. Just now it does not really matter whether we use model, picture, or schema, as long as we remember that we are using them all as metaphorical aids to thinking pictorially about something that is not a picture but an occurrence—the interaction that takes place between the human nervous system and that part of the non-nervous system from which it receives the messages.

In this chapter we have tried to elaborate, for the human case, mostly the first half of the formula for animal knowledge at its widest application—we are still in the "mapping room" between the stimulus and the response—and I have proposed that mapping (or, in more concrete cases, picturing) is precisely the activity that illustrates best the nature of the combination of correspondence and noncorrespondence that obtains between the two parts of reality. We cannot elaborate on the second half of the formula until we have examined the nature of human action, limitation, and possibility, and the problem of values for the human case, which we do in the next chapter.

VI

In preparation for the next step, I should like to present a little excursus on Ordinary Language, considered as a preliminary

nary, fallible, but correctible, *verbal mapping* of the common-sense world, to see if the mapping model of the correspondence might not throw some light on the great variety, yet the general intertranslatability, of the actual languages of the world. Man is often congratulated on being the language-using animal, as if this settled some epistemological problems, but when we take a closer look at the Tower of Babel he has erected out of his inventiveness in this direction, we could easily regret such premature optimism. (In the Bible story, God is driven to the strategem of the confusion of tongues to prevent His creature from reaching overweening heights of power, in the days when the family of man all spoke one language.) In Section II, we said that characteristic features of all maps are: the scale, the grid, the projection, the content, and the omissions; and the projection involves the introduction of systematic distortion, and the content and deliberate omissions introduce selective distortion. We also noted that there are no *ideal* projections, since they are all by definition systematic distortions produced by projecting the "natural" global grid onto a flat surface in various ways, and they are all convertible one into another. As for their desirability, it is only a question of selecting the one that provides the most tolerable distortion for conveying the information for which that particular map is designed.

There are something like 130 languages spoken by more than one million people, each. When we consider all languages, including "dead" languages and some obscure South American and African languages not yet properly catalogued, the *Encyclopaedia Britannica* tells us that there are "anywhere from 2,500 to 5,000 languages in the world (depending on what is counted as a language and not counting dialects that are clearly just that)." Let us now try the experiment of thinking about the phonetics, the vocabulary (in its parts-of-speech rather than its content aspect), the grammar, and the syntax, as constituting the elements of the "grid," the conventional frame of reference onto which the nonlinguistic reality is in each case "projected." The homeomorphous relation that exists between each language and the nonlinguistic reality of which it is a "map" is possible because in each case two sets of categories are being matched—the words, a set of speech categories; and the things and events, a set of nonlinguistic categories. What, then, about the homeomorphous relation between different languages? *Equality* is hardly to be expected—we are not dealing with a case of things equal to the same thing being equal to each other. First, the "same thing," that is, the categorizable nonlinguistic world, allows itself to be categorized in many different ways, as we have seen in the previous section; second, the original relation between each language and the categorizable world is already "warped" from being projected with the systematic distortion introduced by its own rules of vocabulary, grammar, and syntax. Yet, in spite of this, some resemblances must remain, for we would not consider an utterly untranslatable set of words language at all, but gibberish. There must be something in the speech situation that acts as a check both on a too-idiosyncratic categorization of the categorizable world, and on a too-distorted conventional frame of reference or "projection." This something is the need, whether dire necessity or spontaneous desire, of people to communicate with one another, using all the devices of empathic imagining and expressive imagining we described in Chapter Three, including gesture, facial expressions, and actions, and including as a heritage of the speech community's inventiveness the many different ways of saying the same thing.

Even within the same language it is possible to say the same thing in many different ways. (The "projection" is flexible, even though systematic.) Otto Jesperson gave the following example:

He moved astonishingly fast.
He moved with astonishing rapidity.
His movements were astonishingly rapid.

His rapid movements astonished us.
His movements astonished us by their rapidity.
The rapidity of his movements was astonishing.
The rapidity with which he moved astonished us.
He astonished us by moving rapidly.
He astonished us by his rapid movements.
He astonished us by the rapidity of his movements. (20)

As between two different languages, the idea that it is possible to say the same thing in many different ways apparently causes no difficulty to children brought up in a bilingual family. They automatically translate from one language to the other when they carry messages between the parents; for as soon as they learn a word they also learn that it belongs to the set of words with which Papa operates or to the set with which Mama operates. The problem exercises the brains of students of linguistics in no small degree, however, and as would be expected, divides them into opposite camps. Purists among them leap to the defense of the uniqueness of verbal expressions and insist that if a thing has to be said in a different way, it is no longer the "same thing"; that there are really no synonyms; and that there is really no such thing as translation, only paraphrasing in the case of prose, and a new verbal recomposition, or else caricaturing, in the case of poetry. Logicians, on the other hand, point out that all languages are to some degree referential and therefore must have some degree of resemblance, since at the very least one can point to objects or pantomine actions in order to learn their names in one language or another. Furthermore, although we may concede that phonetics represents a largely arbitrary aspect of the structure of languages (even onomatopoeic words are rendered differently in different languages) not only vocabulary, but also the rules of grammar and syntax "stand for" something in the nonlinguistic reality and are therefore referential to that degree and form a basis for establishing resemblances between languages. Jespersen puts the case thus, after listing the most comprehensive grammatical and syntactical categories (parts of speech or word classes, number, cases, tenses, moods, voices, persons, genders):

> Now, some of the categories enumerated above bear evident relations to something that is found in the sphere of things: thus the grammatical category of number evidently corresponds to the distinction found in the outside world between "one" and "more than one"; to account for the various grammatical tenses, present, imperfect, etc., one must refer to the outside notion of "time"; the difference between the three grammatical persons corresponds to the natural distinction between the speaker, the person spoken to, and something outside of both. In some of the other categories the correspondence with something outside the sphere of speech is not so obvious, and it may be that those writers who want to establish such correspondence, who think for instance, that the grammatical distinction between substantive and adjective corresponds to an external distinction between substance and quality, or who try to establish a "logical" system of cases or moods, are under a fundamental delusion . . . .The outside world, as reflected in the human mind, is extremely complicated, and it is not to be expected that men should always have stumbled upon the simplest and most precise way of denoting the myriads of phenomena and the manifold relations between them that call for communication. The correspondence between external and grammatical categories is therefore never complete, and we find the most curious and unexpected overlappings and intersections everywhere. (21)

This admirably moderate statement of the situation is quite in keeping with the map-projection view of the relation between the grammatical structure of language and the nonlinguistic reality, since it allows for a degree of correspondence and for a degree of noncorrespondence, different for each language or family of languages, but systematic within each language. Such a view should help us to steer a middle course between the extremes of linguistic relativism on the one hand

and the ideal of a perfect, universal, or logical language on the other. According to the extreme linguistic relativism of Benjamin Lee Whorf, any language, but in an especially fateful way the one that happens to be our mother tongue, enshrines in its utterly taken-for-granted grammatical structure, a whole world view, a way of seeing things, a "petrified philosophy" (Max Möller), which predetermines the speech community's way of *thinking,* not merely its way of communicating. Thus is established a self-perpetuating "mold" into which the minds of the children are poured in every generation, and from which they can never entirely escape, no matter how sophisticated or cosmopolitan they may become as adults. As Whorf himself puts it:

> We cut nature up, organize it into concepts, and ascribe significances as we do, largely because we are parties to an agreement to organize it in this way—an agreement that holds throughout our speech community and is codified in the patterns of our language. The agreement is, of course, an implicit and unstated one, *but its terms are absolutely obligatory:* we cannot talk at all except by subscribing to the organization and classification of data which the agreement decrees. . . . We are thus introduced to a new principle of relativity, which holds that all observers are not led by the same physical evidence to the same picture of the universe, unless their linguistic backgrounds are similar, or can in some way be calibrated.(22)

Whorf then explains the unanimity in the description of the world by the modern community of scientists by noting that they were mostly reared in modern European languages, with Latin and Greek thrown in for good measure, and these are all dialects of the Indo-European family.

But it must be emphasized that "all modern Indo-European-speaking observers" is not the same thing as "all observers." That modern Chinese or Turkish scientists describe the world in the same terms as Western scientists means, of course, only that they have taken over bodily the entire Western system of rationalizations, not that they have corroborated that system from their native posts of observation. When Semitic, Chinese, Tibetan, or African languages are contrasted with our own, the divergence in analysis of the world becomes apparent; and when we bring in the native languages of the Americas, where speech communities for many millenniums have gone their ways independently of each other and of the Old World, the fact that languages dissect nature in many different ways becomes patent. The relativity of all conceptual systems, ours included, and their dependence upon language stand revealed. (23)

From the standpoint of the map-projection model of the grammatical structure of languages here recommended, the only question is, how far has Whorf overstated the case? It is possible to make map projections in which the systematic distortion is so great that the natural contours of the land become virtually unrecognizable; but what would be the incentive to do so? The natural contours of the land act as a check on the tolerability of the systematic distortion of a given projection. Similarly, the "natural contours" of the sensory world, and especially of the visual world, act as a check on the systematic distortion introduced into each language by its vocabulary, grammar, and syntax. As Roger Brown well puts it (although the term "homeomorphic" would have been more accurate than "isomorphic"):

It is evidently possible for non-linguistic reality to serve as a guide to the categorization of speech. The isomorphic relationship can be useful in either direction. An inescapable visual difference leads us to look for a speech difference. We may suppose that speech has less intrinsic importance than non-linguistic reality, and it is therefore customary to describe speech as a map and the rest of the world as the region mapped. It is clear, however, that when a dirt road turns into a four-lane superhighway we may for the first time notice the difference between a thin and a thick red line on an actual road map. (24)

Thus the more prominent, persistent, fused Gestalts of sensations that we have previously called "objects," as well as the more prominent Gestalts of actions we have called "events," serve as deterrents to a too-idiosyncratic structuring of the visual world. But even more favorable to the intertranslatability of languages are the customs, dictated by the need or will to communicate, of free translation or paraphrasing, of rendering by a phrase in one language what is a word in another, of supplying a mode of describing "in other words" what is named one way in one language, or not at all in another. Roger Brown thus charges Whorf with applying the principle of literal translation unequally:

> In the classroom where the European languages are taught it is customary to assume that psychological processes are fundamentally the same and then to liberalize translation procedures so as to guarantee equivalent meanings. With Shawnee, Nootka, Apache, and Hopi (though never with the European languages), Whorf changed this usual procedure, insisted on a literal translation, and concluded that world views differ. (25)

Even more damaging to Whorf's hypothesis, or indicative of his overstatement of the case, is individual language mutability: each language changes over a period of years; new words are added; new idiomatic expressions are popularly introduced and adopted; poetic "license" is incorporated; slang is invented. There is a kind of internal pressure against the regularity of a given grammar and syntax which indicates that the community of speakers using it is somewhat less securely wedded to it than the thought-molding theory implies. The case is similar with children: they have to learn a language which is a mapping of the child-world, one that enables them to deal with the contents of the child-world and to get what they want in it; otherwise they could never learn it at all. But this does not mean they are wedded for life to the concepts derived from the child-world picture of the nonlinguistic reality.

No doubt there are semantic hangovers from childhood in the later linguistic understanding of every adult—such as animism, personification, word magic (incantation), and the reification of concepts (thinking of abstract nouns as the names of "things"), but we all have to achieve "sophistication" out of these thought habits as best we can, since there is no other way to start. An early training in several languages is probably the fastest way to learn to distinguish the message from the code, to separate the "same thing" from the different ways it can be said. (The invention and relative workability of electronic translating machines would seem to indicate that the trick can be done "mechanically," if nouns, verbs, prepositions, etc., in one language are allowed to do the jobs of other parts of speech and grammatical forms in another.)

Thus the map-projection model for the correspondence between the vocabulary, grammar, and syntax of a given language and the nonlinguistic reality goes part of the way with Whorf in admitting to a systematic distortion inherent in the very structure of language. However, it would also insist that there are ways of getting from one projection to another without losing one's way as far as the major contours of the country or the more prominent landmarks are concerned. By the same token, the map-projection model would not hold out much hope for the ideal of a universal, logically perfect, vocabulary, grammar, and syntax, corresponding in theory to the "real" structure of the world. This is like asking whether there is not a "natural" grid for that structure, as the longitude–latitude system seems to be a "natural" grid for the global structure of the earth, one on which the contours and landmarks would appear in their *undistorted* form. Here, from all we have learned about the multiple categorizing activity of the brain and the multiple ways in which the world can be categorized, the answer would have to be a firm *no*—there is no such "natural" grid for the nonlinguistic reality that is being mapped. In fact, even the longitude–latitude system is not the only possible way of making a squared-off referential system on a globe: any great circle could serve as an equator. Such a system would not distort the contours, but it would not be very useful. Thus the "natural" grid really means the easiest, the most workable grid—and something of the meaning of "the easiest and the most workable" also attaches to the so-called natural categorizations of the world, in terms of which we eventually come to describe the "natural" world, as the self-styled nominalistic authors of *A Study of Thinking* remind us:

> Stevens (1936) sums up the contemporary nominalism in these terms: "Nowadays we concede that the purpose of science is to invent workable descriptions of the universe. Workable by whom? By us. We invent logical systems such as logic and mathematics whose terms are used to denote discriminable aspects of nature and with these systems we formulate descriptions of the world as we see it and according to our convenience. We work in this fashion because there is no other way for us to work." Because the study of these acts of invention is within the competence of the psychologist, Stevens calls psychology "the propadeutic science." (26)

A universal language that would be universal only in the sense that everybody would use it (probably as an auxiliary language) would be another such invention, and several have been attempted, including Esperanto and Interlingua. But such a language would be no more perfectly logical than any of the others, and since it would have a vocabulary, grammar, and syntax, it would have its own systematic distortion built into it, just like the others. It would probably be judged good or bad on the grounds of easiness and workability, and if actually put to long use, it would undergo change over a period of years, and develop its own idioms, dialects, and slang.

So far we have considered only the grid and the projections as characteristics of maps that are helpful in understanding the combination of the correspondence and noncorrespondence between language and the nonlinguistic reality, and in seeing how the great variety of languages does not preclude their practical intertranslatability. It remains to look briefly at the three other characteristics of maps: the scale, the content, and the omissions. I leave the scale to the end, since it represents the limiting case of the applicability of the model.

Of course vocabulary, in its lexical, not its grammatical aspect, would be the obvious analogy to the symbolic content of a map—but we must immediately add that dictionaries give not only the referential meanings of a word, its denotations, and connotations, but also its functions and established usages, illustrated by paradigm sentences in which they are displayed. The vocabulary that one person uses in communication with another is not an impartial inventory of the items in the experienced situation. Rather, it is *a way of calling attention to* parts or features of the situation the speaker considers important, and therefore, also *a way of calling attention away* from those he considers unimportant for the moment—just as the symbolic content of a map exaggerates some features at the expense of others, as a way of calling attention to certain ones for that map's special purposes. Both in calling-attention-to and in calling-attention-away-from, a person's vocabulary is an example of the ground-to-figure relationship which the gestalt psychologists have pointed out is an essential part of the perception of pattern. It is also the basis of the intelligibility of sentences as much as of maps: it is impossible to put everything (or even too much) in one sentence and still make sense, just as it is impossible to put everything into one map and still have a map instead of chaos. Selective distortion and its complement, omission, are as much an inherent part of the language in relation to the purposes of its users as they are of maps in relation to the purposes of their users. The language user, just to make sense, is forced by the very nature of language to thematize his experience according to the principle: important/unimportant.

In the field of formal education, learning new subject matter certainly in part involves learning the vocabulary that has been developed for the curriculum's special needs, and we would not say a lawyer or a doctor knew his subject if he could not handle its technical words. When anyone learns a new subject, he acquires, as it were, a blown-up map of a local region. On this map—over many years perhaps, and with the assistance of some or all of the forms of imagining we distinguished in Chapter Three—special "symbolic devices" have been invented to discriminate and communicate features that could not appear on more inclusive or generalized maps. Most of our academic knowledge consists of a wide assortment of such maps—how wide and how assorted and how sometimes apparently unrelated, can be seen by reading the course listings in any college catalogue. With the assistance of these maps as well as the preliminary "common-sense" mapping known as Ordinary Language given to us by our culture, we operate in our matrix world, receiving messages, sorting them out according to the cognitive maps on which they belong, correcting and extending our maps, and planning our responsive actions according to the best maps we already possess. It is no wonder that, the more we know, the less it seems credible that all these maps could ever be combined and overlapped to form a single global map of the world—but this is to anticipate, for such a puzzlement is nothing less than "the unique problem of totality," which we take up in Chapter Seven.

All models and constructs have limited applicability, otherwise omniscience would be in principle achievable; and the mapping model of language is no exception. Its limitations are most obvious when we consider the characteristic of all maps called *scale*. For whereas every map contains in the "key" an exact proportionate relationship between its area and the actual area of the earth's surface, we ask in vain for the diameter, circumference, or total area of anyone's matrix world. Still, with a certain looseness of interpretation, we can attach some meaning even to the concept of scale for language. We would surely agree that a child operates with a smaller-scale map than an adult, an ignorant person with a smaller-scale map than a well-educated individual. Other things being equal, more detailed knowledge can be conveyed by the same symbolic devices on a large-scale map than on a small-scale chart. Since the analog of the area of the earth's surface in this model of language is the total world of human experience as explored and mapped thus far (i.e., as expressed in words), a particular language could be considered as large-scale or small-scale in some parts rather than in others, according to the degree of thoroughness with which certain "areas" of human experience had been explored by it. In point of fact, different languages, or at least language families, do differ with respect to the degree of differentiation and symbolic representation (verbal accuracy) with which they have explored those areas of experience their cultures have found the most significant, either for the preservation of life or for the enlargement of consciousness. Thus not only inventors, scientists, and philosophers, but the makers of literature also, are the Columbuses, Magellans, and Vasco da Gamas of language; and, like Columbus, they must be willing to set out upon their voyages with only the most fragmentary and inadequate maps of the unknown regions, not always finding what they were looking for and sometimes discovering what they did not expect was there at all.

REFERENCES

1. Erwin Raisz, *General Cartography,* McGraw-Hill Book Company, New York and London, 1938, p. 1.
2. Raisz, *General Cartography,* p. 2.
3. Raisz, *General Cartography,* p. 2.
4. Raisz, *General Cartography,* p. 118.

5. Wiener, *The Human Use of Human Beings,* p. 48.
6. Wiener, *The Human Use of Human Beings,* p. 33.
7. Th. Dobzhansky, "Evolution—Organic and Super-Organic," *Bulletin of the Atomic Scientists,* May, 1964, p. 5. Reprinted by permission of *Science and Public Affairs, the Bulletin of the Atomic Scientists.* Copyright ©1964 by the Educational Foundation for Nuclear Science. Adapted from The Rockefeller Institute Review, April, 1963, Published by the Rockefeller Institute, New York.
8. Bruner et al., *A Study of Thinking,* p. vii.
9. Bruner et al., *A Study of Thinking,* p. 1.
10. Bruner et al., *A Study of Thinking,* p. 2.
11. Bruner et al., *A Study of Thinking,* p. 2.
12. Jean Piaget, quoted in Bruner et al., *A Study of Thinking,* p. 3.
13. Bruner et al., *A Study of Thinking,* p. 2.
14. Bruner et al., *A Study of Thinking,* p. 4.
15. Baensch, "Art and Feeling," p. 15.
16. Bruner et al., *A Study of Thinking,* p. 5.
17. Bruner et al., *A Study of Thinking,* p. 6.
18. Bruner et al., *A Study of Thinking,* pp. 12–13 passim.
19. Raisz, *General Cartography,* p. 2.
20. Otto Jespersen, *The Philosophy of Grammar,* George Allen and Unwin, Ltd., London, 1925, p. 91.
21. Jespersen, *The Philosophy of Grammar,* p. 54.
22. Benjamin Lee Whorf, *Language, Thought, and Reality,* reprinted by permission of The M.I.T. Press, Cambridge, Massachusetts, 1956, pp. 213–214. Copyright ©1956 by the Massachusetts Institute of Technology.
23. Whorf, *Language, Thought, and Reality,* p. 214.
24. Roger Brown, *Words and Things,* The Free Press of Glencoe, Glencoe, Ill., 1958, p. 216.
25. Brown, *Words and Things,* p. 231.
26. Bruner et al., *A Study of Thinking,* p. 7.

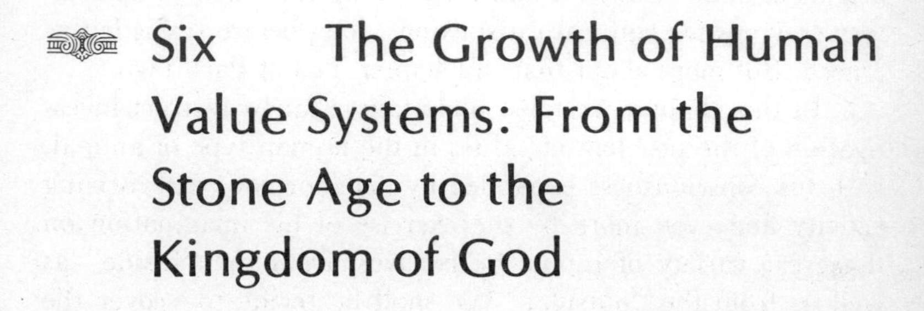

Six The Growth of Human Value Systems: From the Stone Age to the Kingdom of God

I

When we left the animal value system in Chapter Four, we had just reached the point of considering the peculiarities of the specifically human type of childhood development. We noted the severity of the human child's helplessness (not being able to move about or perform for a long period of time, as compared with a calf or colt, which can do so within hours of birth) and the sheer length of his period of dependency on grownups.

As we noted in Chapter Four, when the human child has reached the age of such physical independence that he might conceivably survive on his own as a large-brained animal (4–5 years?), he has been introduced by the very agency that had kept him alive up to that point—the societal factor—into a world of grownup's values and disvalues about which the ordinary animal knows nothing, and to a whole range of possibilities and limitations that the ordinary animal need never face.

This "education" is justified in making the child "fit" into his society, but there are negative side-effects. Anyone who has ever heard a stupid parent threatening and reviling a 3-year-old, to "make him behave," need speculate no further on the origins of adult neuroses, inferiority complexes, and compensatory cravings for cultural satisfactions, long before sex is felt as a need. But more about that in Chapter Ten of Part Two.

In this chapter we must undertake a fairly abstract investigation of the problem of values in the human type of animal, with his consciousness expanded by his enormous categorizing activity and even more by the exercise of his imagination on the great variety of inputs he receives, from the "inside" as well as from the "outside." We shall be trying to "cover the ground" ordinarily found in theory of value, ethics, social psychology, and cultural anthropology. The abstractness is regrettable, since we now know that in the human context, abstractions purchase their generality at the expense of "leaving out" the convincing, experienced concreteness. But we do not have time or space here for a historical survey of actual cultures from the Stone Age to the present and the expected future.

Let us begin by nailing down some of the questions generally treated in theory of value:

(1) What is value, as against fact?
(2) Are values cognitive or emotive, rational or arbitrary?
(3) What is the meaning of "ought" as against "is"?
(4) Can you get from an "is" premise to an "ought" conclusion in a syllogism (the naturalistic fallacy)?
(5) Can you get from "is" to "ought" any other way?
(6) Is the language of moral discourse basically descriptive? presciptive? commendatory? ends-and-means relating? goal-defining? society-regulating? law-proclaiming?
(7) Is there only one ideal and absolute set of values for man, or does the multiplicity of value systems demon-

strate their necessary relativity to cultural and historical situations?

(8) Is there an "existential a priori," that is, a specifically human nature, that makes some value systems more appropriate to man than others?

(9) Is it possible to devise metacriteria by means of which the great variety of human value systems could themselves be evaluated, and would this not lead to an infinite regress?

(10) What is the "authority" that gives value systems their coercive power?

In answer to question 1, it must be pointed out from the start that a certain amount of confusion has been introduced into traditional as well as modern discussions of values simply by the reification of the concept, as, for example, in these expressions: "Ah! The eternal *values*!" "You ought to live so as to increase *values*." "The highest *good*." "The greatest *good* of the greatest number." It is almost impossible to avoid this traditional way of talking about values as if they were things, and the custom is sanctioned by the usages of ordinary language. Our evolutionary outlook, however, forces us to realize that *values* first entered the world as the *actions* of mobile forms of life, actions that in themselves constituted the evaluation of situations as beneficial or harmful, so that *evaluation* is one of the *facts* of animal life (i.e., it is impossible for animals to live nonevaluatively). Another way of putting this is to say that for all animals, including that most sophisticated and supposedly impartial animal, the rational man, every action is a cryptoevaluation, and every evaluation is a cryptodirective for action. It is a delusion promoted by the objectivizing tendency of this overintellectualized animal that man lives (exists) in a world of "facts," to which "values" are somehow mysteriously (or arbitrarily or subjectively or whimsically) added, *ex post facto*. The truth is rather the other way around: the first "facts" that entered the categorizing activity of animals (the cognitive or intellectual activity of animals) were the sensory-perceptual items that appeared on the animals' mapping of their environment in terms of benefit expectancies and harm expectancies. How should we, at this point, answer the first question? Let us say that, in the most primordial sense, *a value is a fact categorized as harmful or beneficial by some living creature.* This view would be in keeping with the second half of our general formula, namely, that the "encoding" must be in such a manner that the creature's future performance can be either perpetuated or modified by its past performance. The animal's preliminary classification of certain "facts" as harmful or beneficial must be confirmable or disconfirmable by further action, and there must be some possibility of reclassification, if knowledge is to take place. And, as we have seen, it is the inherited primary emotion repertoire that functions as a guide, both for the preliminary classification and for any subsequent reclassification based on experience.

Under these circumstances it is difficult to see why, as question 2 seems to imply, cognitive and emotive meaning in relation to values could ever appear as dichotomous, or even as separable, since both sensory perception and emotive guidance are necessary to the *cognition of values* (i.e., the categorization of facts as beneficial or harmful). However, we did notice two conditions under which the emotions seemed to lose their preservative function even for the animal, namely, when they occurred at too low intensity, or too high intensity. In the first case there was a loss of attention and guidance discrimination; in the second case there was a general lowering of perceptual discrimination because of too much physiological involvement. Of course the human case is more complex, but it may be that the opposition, cognitive versus emotive, has its origin in some similar circumstance of human behavior.

The opposition rational versus arbitrary, or whimsical, raises more questions about the use of the term "rational" than we have time to consider. Man is an animal, and animals cannot live and survive nonevaluatively. Is a man who is behaving self-preservatively, using all the primary emotions correctly to guide his actions toward benefits and away from dangers—is such a man behaving irrationally? Arbitrarily? Whimsically? Surely if he were on a desert island we should all applaud him as a paragon of rationality. It is only when the self-preservation of one man conflicts with that of another that "ethics" raises its ugly head, and the traditional name-calling begins. If we regard ethics in its primordial function as *the minimum traffic rules of society*—the very least individual self-restraint that makes the benefits of life in society possible—we can easily see that a law-abiding citizen will come to be regarded as a rational animal, though precisely at the expense of "doing what comes naturally." This is what makes "naturalistic ethics" so peculiar, since ethical man is already an unnatural animal, in fact, a supernatural or a beyond-natural animal. But of course historically, tribalistically, no such reasoning took place, because just to get people to behave in this nonnatural, self-restraining, tribally promoting way, early man had to invoke all kinds of gods and demons, promises and threats, rewards and punishments, tribal destiny and glory, and so on and so forth. These considerations constitute an indirect answer to question 6, namely, that all these kinds of discourse, and even more, will be needed in this enterprise.

As for getting from "is" to "ought" in a syllogism, let us just try it. A syllogism is the spelling-out of the relations between a class, its criterial attributes, and the members of the class. There exists a class of beings called man, and one of man's attributes is mortality. Socrates is a member of this class. Therefore Socrates is mortal (or he does not belong to the class). What have we learned that we did not already know if indeed "*all* men are mortal" is the case? Nothing. Therefore, if we want to get an "ought" conclusion we must start with an "ought" major premise. All men ought to be honest. Socrates is a man. Therefore Socrates ought to be honest. "Is" premises need to be verified or disverified; "ought" premises need to be defended or opposed. Here is the proper place for moral idealism: providing reasoned defenses for such "ought" premises about man as consequences to society, tribal survival, individual ideals, long-range benefits and social goals, and the will of God in a specific religion—always remembering that an "ought" premise is an evaluation, and an evaluation is a cryptodirective for action.

This brings us directly to question 7, the problem of the "relativeness" or "absoluteness" of values, especially when these terms are used in a pejorative or commendatory sense about human values. If values are facts categorized as harmful or beneficial by some living creatures, it is obvious that they must be species-specific, that is, "relative" to a particular creature's environmental niche and his receiving–encoding equipment. What is categorized as harmful or beneficial by a bird will not be so categorized by a fish, and certainly what does not appear in the environment of either species cannot be evaluated by members of that species at all. But in spite of the species-specific nature of the great variety of actual animal value systems, they all have the following trait in common: each is an animal's response to certain prototypical situations that repeatedly confront him in carrying out his task of staying alive. Assuming that *Homo sapiens* is a species, it would seem that human value systems, no matter how varied in detail, must also in some sense be species-specific. They must be "relative" to the human environment and the human receiving–encoding equipment, precisely in order not to be arbitrary or whimsical, but in order to function *at least* preservatively. The "relativeness" of values to historical and cultural situations should not be interpreted as demonstrating "the arbitrariness of values," but just the opposite—as showing that a value system, a set of interrelated evaluations of facts as harmful or beneficial, can be "reasonable" only relative to a specific cultural niche. In that case, what does the commendatory term "absolute" or "ideal" mean in regard to values? We know of no animal or human being that lives in an absolute or ideal environment and is equipped with absolute or ideal receiving–encoding equipment—and such a super-animal, if we could even imagine it, would surely need to do no evaluating at all. Those philosophers who try to use the terms absolute or ideal in a commendatory sense—Platonists and idealists, generally—forget that limitation is of the very nature of evaluation, and that to such a Splendid Beast, one who had infinite possibility and no limitation (some people's idea of God), value could have no meaning and even his relationships to evaluating beings would be unimaginable. The theory of ideal or absolute values, then, aside from being a contradiction in terms, is a disguised rejection of the natural limitations of man; but when adumbrated in a less fantastical way, as the theory of the "highest good" (i.e., something that *ought* to be evaluated as most beneficial in a specifically human value system), it is perhaps an attempt to provide an answer to questions 8 and 9.

II

In this section we must try to find affirmative answers to questions 8 and 9. Having received our clues from the animal kingdom that (*a*) animals cannot live nonevaluatively, *(b)* their evaluations, in order to function preservatively, must be species-specific, and *(c)* knowledge of values consists of the mapping of the environment in terms of benefit expectancies and harm expectancies, our strategy must be to show that yes, there is a species-specific human nature. Then we are obliged to devise metacriteria by which to judge actual human value systems to determine the degree to which they are species-specific and therefore can function preservatively—in the widest *human* sense of preservative, not in the merely animal sense.

Let us now briefly recapitulate the chief differences we have learned so far in this essay, between the kinds of worlds the various animals live in and the kind of world that *Homo sapiens* lives in.

First, there is man's oversized brain. On the one hand it gives him the survival advantage of being a superclassifying machine in a classifiable world; but on the other hand it gives him the disadvantage of so long a period of childhood helplessness that some form of societal nurture and protection becomes mandatory for his survival. This is the overriding anatomical, biologically inherited difference between *Homo sapiens* and even his closest cousins among the animals. The chimpanzee is a grownup at the age of 5 years, able to get his own food, to compete with others, to run from danger, to defend himself, and to seek a mate. And where is the human child at the age of 5? That depends on how his culture, and even his class status within that culture, has nurtured him, physically and psychologically, up to that point. We (Americans, e.g.) live in a culture that permits cats, dogs, ponies, and even farm animals to receive better babyhood treatment than millions of our human children actually receive. Those who survive this early obstacle course may have proved that they are *physically* the "fittest"; what it has done to them psychologically, however, is simply incalculable in terms of our present knowledge. But we can safely risk the wager that it does not make them *psychologically* the fittest for a good grownup human life. Don't we want something better for our own children? This example is an illustration of how it is possible to get from "is" to "ought" by means other than a syllogism (question 5) and also how a specific human nature will yield metacriteria for a humanly adequate value system.

Psychologically, of course, we begin with the primary emotion repertoire, delayed by the early helplessness to the point where the consciousness in which it can actually perform the survival tasks is already enormously enlarged compared with the animals' consciousness. I mean things like memory, anticipation, curiosity, verbal expertise, pride in performance, and self-image,—all these change the child from the "cute doll" of babyhood to the "little monster" of age 5 to 10.

Next, there is man's visual receiving-encoding equipment, which bombards him with more discriminable bits of information than he knows what to do with (whenever, out of curiosity, he chooses to pay attention to them). Of course there are plenty of animals with curiosity and sharp eyes, but the difference seems to be one of degree and the length of the human attention span. Animals (and children) are too easily distracted. And then there is the overdeveloped functioning of man's imagination, whose several operations for enhancing knowledge, as well as for distorting reality, we anatomized in Chapter Three. This "sudden" evolutionary expansion and proliferation of the brain created for man a different world, which we called the *matrix world*, a space-time *lebensraum* extended by imagination from the perceivable and classifiable world into what seemed to be limitless reaches of space and time, and evoking in man a tension between the threat of chaos and the hope of order. At a hint from cybernetics we noted that such is the way the world must look to one who represents, on the most conscious level, the local enclave of the antientropic tendency in the universe, including feelings of disorganization, dissolution, and death, associated with chaos, and feelings of life, growth, and fulfillment, associated with order. The threatening-promising aspect of the double imensity of space and time that is distinctly man's world as against the other animals', is what is experienced by him in certain moments as the *mysterium tremendum et fascinans* and in certain other moments as the *philosophical wonder* with which all metaphysical speculation begins—first, the wonder that there should be anything at all, when there might have been nothing; and second, that it should be of this nature, when it might have been of some other. Both these experiences lead to the question: *what is it*, then, in its totality? And what is its import for me? We must leave these questions for the next chapter, since they involve the unique problem of totality. But even here, where we are considering only the evaluating activity of man, we cannot escape from the influence on man's thinking of his consciousness (however dim or intermittent or inchoate) of the *presence* of the mysterious totality as the constant background of all his busy actions and evaluations in the foreground.

In order to have something definite before our eyes, let me attempt a model or crude map of the *matrix world* (Figure 1) to schematize some of the relationships in it we have discussed so far. In this model, it is again unnecessary to make any "sharp cuts" between man and the other animals, for their consciousness would be represented by the same schema in extremely contracted form with regard to the degree of presence of the present, past, and future in it. Nevertheless, since they are able to learn from their experience to some extent, we must credit them with *some* degree or kind of memory and anticipation. In this contracted form, then, we might suppose that the matrix worlds of anthropoid apes, of very young children (before performance becomes a problem), and of grownups who are severely retarded or brain-damaged, would not differ too greatly from each other.

This model can be asymetrically expanded or contracted like a bubble* and can therefore represent the differences, noted in Chapter Three, between the matrix worlds of children and grownups; among grownups, between the matrix worlds of the educated and the uneducated; and among the educated, between the matrix worlds of those who have used their perception, categorization, and imagination to expand

* Physically, the "globe" in Figure 1*a* is more like the star maps projected in a planetarium, since the map user is pictured on the interior of a sphere; but we also have celestial globes that are looked at from the outside. Thanks to the imagination, it can be looked at either way.

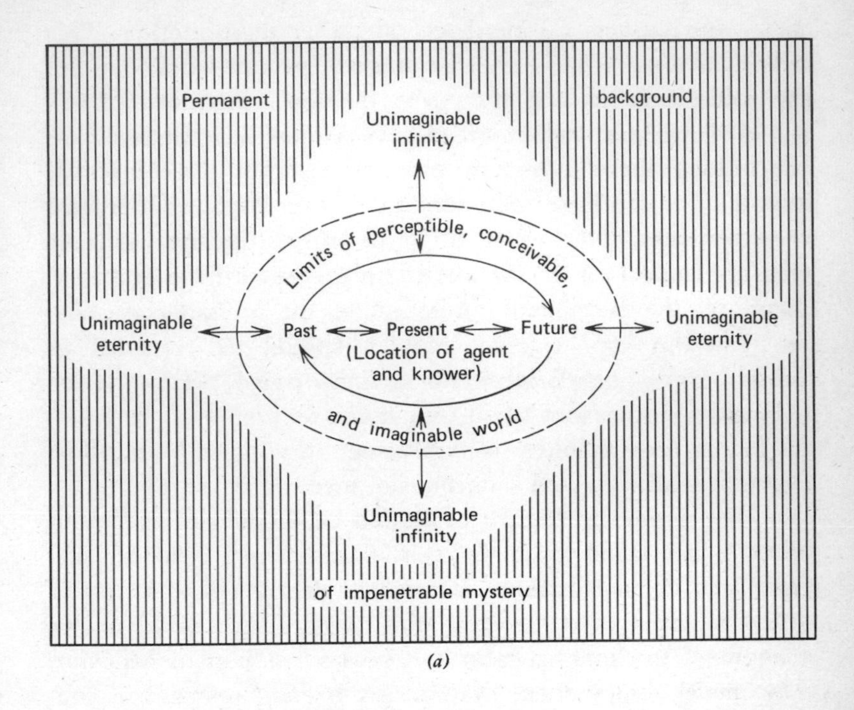

(a)

Beneficial/Harmful means:

For the past:	For the present:	For the future:
Redemptive/ Enslaving	*Preservative/ Destructive*	*Self-Transcending/ Self-Arresting*

Figure 1 *(a)* Crude map or model of the matrix world (the human type of consciousness). *(b)* Existential metacriteria for expanding the animal criterion of action—beneficial/harmful—to fit the expanded consciousness of *Homo sapiens.*

their consciousness in one direction rather than another. This tells us at once what a "well-rounded" person would be, but the model tells us also that even the most expanded matrix world of the most well-rounded person must have limits, if we are talking about a human creature and not some fantastic animal. For we have seen that there are costs and burdens, even penalties, attached to this type of consciousness that no creature could bear or support in an unlimited form and that, therefore, the permanent limitation of the human consciousness by the impenetrable mystery should be regarded as "good" (i.e., beneficial) and acknowledged with thanks. ("Human kind cannot stand very much reality."—T. S. Eliot) As far as accountability is concerned, this "good" limitation exonerates all sorts and conditions of men.

The arrows in Figure 1*a* should be thought of as "influences"; they stand for both positive and negative factors that enter into decision making, as well as degree of awareness or attention given in the several directions. They should not be thought of too simplistically as "causes" of human behavior. The model shows, then, that in any *present* moment of consciousness of the grownup human type, the past and the future are *co-present* in various ways. For example, as an extreme case, in the consciousness of the absent-minded professor, of the remorse-stricken person, or of the daydreamer, the past or the future may be strongly present, by way of imagination, thus blotting out the consciousness of the perceivable present.

It will be remembered from our study in Chapter Four of the animal value system that their criterion of action—beneficial/harmful—meant preservative/destructive (for the animals, mostly pleasant/unpleasant except as modified by experience), and that it arose out of the very nature of animal existence (as against plant and inanimate existence) regarded as an evolutionary attempt to establish mobile forms of life in the medium of the immobile and the lifeless. It was a sort of positive–negative guidance system to promote animal survival. The question now is, how would such a guidance system have to be expanded and modified to serve the needs of the expanded human type of consciousness which includes the past and the future in various ways? It seems obvious that the criterion: beneficial/harmful would have to be applied to the past and to the future as well as to the present, since any two of these, as the arrows in the crude map show, can influence the third, and all three are usually operating simultaneously. That is to say, in the "mapping room" between stimulus and response, among the many maps to be consulted in deciding on a strategy of action, are whatever maps are available, in that culture, of the past and future in terms of benefit expectancies and harm expectancies. The maps may be crude, full of errors of fact and errors of evaluation, but they will be consulted anyway, since animals cannot exist nonevaluatively.

III

By the time we reach question 8 and start to talk about value systems being appropriate to the specifically beyond-animal, beyond-child quality of human nature, we are already impinging on the problem of the wholeness of man, that is, the degree of inclusiveness of his consciousness in the different stages of his individual growth from birth to death. We cannot answer questions 9 and 10 about the metacriteria and the authority for value systems, without betraying, or confessing, our own expectations of the human animal from the standpoint of his wholeness, his completeness—the most he is capable of rather than the least. And we cannot do that without wandering into some of the territory more properly covered in Chapters Seven and Twelve.

Figure 1*b*, the metacriteria in the crude map of the human type of consciousness (Figure 1*a*), represents the expansion of the animal action criterion: beneficial/harmful to the past and the future, derived from my understanding of the biblical concept of the wholeness of man. Readers who disagree with this concept must substitute their own wholeness concept, being as specific as possible in spelling out their disagreement regarding how the criteria for past-present-future must be changed for grownups. (They might want to consult Appendix Two to see how I came to prefer the biblical view of man to others, since I was certainly not "brought up" on it.) Unfortunately for making comparisons, the Bible does not contain any handy abstract of its concept of man, in either philosophical or psychological terms, but rather is itself an *embodiment* of that concept in dramatic, mythic, legendary, pictorial, and visionary terms. (The same is true of the Koran, the Bhagavad-Gita, etc., in relation to their wholeness concepts of man.)

But since this religious way of talking about the religious subject matter, God and man, is supposed to be gobbledygook to modern ears, let us do a quick exegesis of Psalm 103, to see if we can make sense of it as part of a value system for grownups.

What is being talked about in this psalm, and for what purpose? The most inclusive reality, including all that is transcendent and immanent, all that is past, present, and future, is being talked about, and for the purpose of praise. The Lord is the identity category or proper name that is being applied to this strange reality, and praise is recommended to the human soul for the very fact that this category can be *named*. "Praise the Lord, O my soul, and all that is within me, praise his holy Name." But why praise? Surely this astounding *reality* does not need to *receive* my puny squeaks of praise, and surely it includes much that elicits fear and horror rather than praise. It is my *soul* that needs to *give* praise, in order to "forget not all his benefits". There follows in the psalm a list of praiseworthy benefits, natural and human. They are qualities attributable only to the most ideal of fathers (or mothers), all dramatized and presented with considerable poetic exaggeration: the parental patience, beneficence, feeding, healing, preservation, guidance, righting of wrongs, mercy, forgiveness, knowledge of the child's weakness, wickedness, and his being "but dust." This is the properly mythological way (see Chapter Twelve) of talking about God as immanently revealed in the structures of the world and the possibilities of human nature. *Wherever* these actions and attitudes occur, is the implication, *there* God is acting as the Divine Father, and at the same time "showing" man what is good. These are the qualities that should be encouraged and admired in grownup men and women; the opposite qualities should be discouraged and not admired. If actual fathers and mothers displayed this kind of behavior as the most commonplace "doing what comes naturally," no one would sing its praises nor need to be reminded of its benefits. The very fervor of this psalm, coming out of the social milieu of patriarchal tribalism, leads us to suspect that good fathers were not commonplace and that the need for them was sorely felt. But that is a different story. Suffice it to say that in Psalm 103, God is showing man what is good, and for this should be praised.

Now this "showing" is denominated by theologians as *revelation*, and it is often interpreted in the misleading sense of man's being shown something utterly strange and unheard of, and therefore unrecognizable. It seems unlikely, however, that what is shown in revelation, and can be recognized by man, could be anything other than what has already in some form or degree appeared in human experience but *has not been properly evaluated and cultivated or discouraged* until the time of the revelation. Thus long before there was any monotheism, there must have been some loving fathers, as well as some bad fathers, otherwise the comparison between the One God and a loving father would make no sense either to those who accept it or to those who do not. The same goes for such comparisons as Lord, King, Priest, Governor, Mediator, Teacher, Judge, and Redeemer.* Before monotheism, the protective, nutritive, guiding, healing, teaching functions of the loving father were distributed, somewhat haphazardly and inconsistently, among many gods and goddesses. As for the other God-comparisons, they were often actual heroes or offices, deified. Even if we call them personified abstractions, there is no getting away from the fact that they had *some* "referents" in human experience, good or bad. This accounts for the large overlap of religious themes and ideas in the different religions, even though they do not "all teach the same thing." They all use the available fund of human experience, but they make different mappings of the divine–human situation according to such cartographic principles of selection as: important/unimportant; beneficial/harmful; and abiding/transitory. People who are converted from one religion to another on the basis of truth, rather than, say, on the basis of convenience, sentimental attraction, wishful thinking, or syncretistic tolerance, are converted because they think the religion they are adopting contains a more adequate mapping of the divine–human situation than the one they are leaving—and what else can they do? But here

* These God-comparisons are not easily fitted to one another in any rationalistic schema, and some of them can be quite dangerous to the religion that recommends them, when taken literally by a superstitious people, that is to say, the usual pious masses. When Constantine was converted to Christianity, the only kind of kingship he could "understand" was benevolent despotism ruling an empire, and the character of the Church he spread around the Mediterranean increasingly took on the forms and powers and ideologies of Constantinism. Kierkegaard thought that the conversion of Constantine was the greatest disaster in the history of Christianity. *Finitum non capax infinitum*, hence absolute silence is the correct response to God and the ever-needed corrective to all God-comparisons. That leaves unanswered the need to make order out of chaos in the finite, which Constantine was doing in his way, for his time, using the all-too-willing Christian religion as the "cement" for holding his empire together. For those who want some form of Constantinism to supply the model for the coming planetary order, Islam would be a better religion to use, the Koran being filled with God-comparisons suggesting a benevolent despot ruling the universe.

the term "adequate" again touches on the unique problem of totality.

At this point, readers who cannot abide any God-talk, or who are "turned off" by institutional religion of any kind, must simply substitute their own evaluations of human nature and human destiny, to see how and to what degree the animal value system must be changed in order to be species-specific for *their* idea of human nature.

We must remember that we are not writing rules for saints or mystics in this chapter. The metacriteria must be universal enough to apply to the entire species from the Stone Age to the present and to the expected future as of the present time; yet they must be sufficiently concrete to allow every human being to identify them as real elements in his own experience, recognizable through empathic imagining in children and grownups, even in cultures quite different from his own.

First, the present. Insofar as, physiologically, we are no different from the other mammals, and our task of staying alive constrains us to perform daily acts promoting life and avoiding death, for *Homo sapiens,* the criterion of action: beneficial/harmful still means largely: preservative/destructive. But the conditions are now different, by at least two of the factors we have already noticed, the childhood helplessness and the societal factor. These two combine to necessitate an early modification of the hereditary emotion repertoire. The parents have to *teach* the child to behave preservatively; that is, they have to modify the emotion repertoire in such a way that items in the pleasure category will become discriminated and reclassified as preservative (beneficial) pleasures and destructive (harmful) pleasures, and items in the pain category will become discriminated and reclassified as preservative (beneficial) pains, and destructive (harmful) pains. In the absence of any ends-and-means reasoning ability in the child, the parents have to use the emotion repertoire itself to accomplish this purpose—especially the love/hate "leverage"—in such forms as coaxing, warning, commanding, punishing, threatening, rewarding, scolding, praising, forbidding, and encouraging. Contrast this with the way one teaches a grownup the criterial attributes by which to distinguish the edible from the poisonous mushrooms, a purely "factual" description easily separated from the ends for which it may be used. No sooner does the child acquire the power to hurt others besides himself than the parents must include in their "do's" and "don'ts" those forms of behavior by means of which the society seeks to preserve *itself*—they must discourage these forms of behavior that the society will not tolerate in a grownup and encourage those which the society will admire and even reward. Perceiving only dimly (or not at all—no social contract!) the connection between the survival of individual societies and the racial survival of this species, actual societies try to allay their own fear of destruction through *solidarity*, arrogating to themselves the right to punish, demand sacrifices of, and reward its members according to their usefulness for this purpose. By this indirect means, through the preservative social ordering compounded of an inextricable mixture of common sense, magic, and myth in its interpretation of past-present-future (possibly what Émile Durkheim and Lucien Levy-Bruhl meant by "the collective unconscious"), awareness of the past and the future is introduced into an individual's mind when he is still a child, largely living in his own present. Under these circumstances, errors of evaluation, such as taboos and commandments, family and cultic practices, mistakenly thought to be beneficial or antidestructive, are smoothly passed from generation to generation, and only a large-scale event, such as conquest or enslavement or religious conversion, is likely to change them. Putting Socrates to death for corrupting the youth of Athens strikes us, at this remove, as an error of evaluation; but to the defensive Athenians, it probably represented an effective means for reducing anxiety over the solidarity of the city-state. Truth, including correct evaluation, is the first casualty of war.

The question, "What do you want to be when you grow up?" is one that the animals, even if they could talk, would never ask of their young. For without making any effort other than to stay alive, the young of all animals other than man grow into the most perfect specimens their heredity and environment will permit, if indeed "perfect" is the word to use here. With the expanded type of consciousness suggested in Figure 1*a* and the enormously increased range of possibilities in the action repertoire it brings about, no such simple relationship between an individual's childhood and maturity can obtain, and still less between the individual's past, present, and future, in any one moment. A specific condition of human as against animal existence is that human becoming involves choice, and with that choice the affirmation and realization of certain possibilities *at the expense of others* (see Robert Frost's *The Road Not Taken*, and W. B. Yeats's *The Choice*) which in that choice are negated perhaps forever, because of the irreversibility of time. Animal becoming involves choice, but only for the sake of survival; beyond that, animal becoming is simply biological maturation from youth to maturity to death.

The self-transcendence of man is a subject up to now explored mostly by poets and existentialist philosophers, and it lacks an appropriate vocabulary. By naturalistic philosophers it tends to be identified simply with growth and development in the biological and psychological sense, and by rationalists and religious idealists it tends to be treated as the appearance and manifestation in man of something strange, a Monad, a Soul, a Spirit, the Reason, that "homunculus," the ghost-in-the-machine, all hypostatized, giving rise to all the insoluble problems of psychophysical dualism. Let us simply call it the specifically human type of growth, noting its relative independence of biological maturation: a man can easily reach the age of 75 years and not achieve as much self-transcendence as a (humanly) precocious teen-ager. Therefore its opposite is *self-arrest*, a nonachievement not to be confused with anything psychobiological like mental retardation.

The earliest form of self-transcendence, both in the species and in the individual, is the gradual discrimination of consciousness into two parts, self and not-self, actor and powers acting on him, subject and object, self and world. How long did this take prehistorically? From Neanderthal to Cro-Magnon? Merely 75,000 years? Or much longer? And was there a preverbal form, such as pointing or other gesturing? In English, we unfortunately call this discrimination self-consciousness, but the narrow meaning of embarrassment is not without significance. The entire repertoire of primary emotions and their simpler mixtures can now be experienced by the acting self as applying to the self and the other-than-self, separately. Let us just try it out with self and world and the primary emotions, using world as other-than-self: anger at self, anger at world; surprise at self, surprise at world; joy in self, joy in world; acceptance of self, acceptance of world; fear of self, fear of world; sorrow over self, sorrow over world; disgust at self, disgust at world; expectancy of self, expectancy of world. Some of these are quite sophisticated emotions, not reached by the human child until long after the world has itself been discriminated much more (e.g., into people and not-people, my family and not-my-family, my tribe and not-my-tribe). I leave it to the reader to run through the same emotion words with regard to these further world discriminations, and to use also the primary dyads, pride, love, curiosity, awe, guilt, misery, cynicism, aggression, with regard to them (see p. 89).

With such complication of the consciousness brought on by self-consciousness, it is scarcely odd to find the need for simplification beginning to show itself again in order to avoid chaos, in this case a chaos that would make goal-directed action out of the question. Accordingly, we find that the simpler emotions—those which are in the medium intensity range and are, therefore, not too weak to be ineffectual toward action and yet not too strong to be impossible to maintain, physiologically, for more than a few minutes (rage, ecstasy, terror, loathing, etc.; rage is said to be impossible to maintain at top intensity for more than 4 minutes, hence: count to ten!)—those emotions tend to settle down into long-range attitudes, to crystallize as prevailing *emotion habits*, and eventually to determine the overall character of the person as traits, dispositions, and temperament. Now let us run through the same emotion words again, this time thinking how, both as self-directed and as world-directed, they describe the emotion habits of various character types we know. It is easy to see how in this manner, the *society's* value system gets precipitated into its citizens' character structure as typical attitudes, dispositions, good-will or ill-will toward one's own kind or strangers, and even as the various so-called *national* temperaments, such as the British, French, Slavic, or Chinese. And of course the same holds for the "typical" religious character structures, such as the Jewish, the Christian, the Buddhist, the Stoical, and the Islamic. Nowadays, we witness the odd phenomenon of many people suffering from what might be called a Christian character hangover: they were brought up by their parents or by more subtle influences in their culture to develop various typical Christian character traits and are now stuck with them, even though they can no longer believe in "all that religious hocus-pocus" which originally grounded such dispositions in theological and ethical convictions. They feel "*ni chair, ni poisson,*" neither Christian nor definitely something else, but baffled, and in a permanent state of identity questioning. Probably the same phenomenon can be observed in modern, "awakening" communist Asia, as a Hindu, or Buddhist, or Taoist, character hangover, in the midst of the rising Maoist culture.

After all, where would the bite of ethnic humor be if there were not some truth in cultural stereotyping? But people are still individuals, and it is even against this character typology, regarded as one's national or religious or cultural destiny, that the self-transcendence of man must be distinguished, for the problem is compounded by both creativity and uniqueness.

Once the separation into self and world has been effected, how does further self-transcendence come about? The worlds of even the most primitive tribes we know today are no simple things, and each is differentiated along certain themes or important recurrent life events that are partly biological, partly psychological, and partly spiritual. The growing child, even past helplessness, has no choice but to "take them in," learn about them, and develop attitudes toward them. These themes are: other selves, our tribe, other tribes, past-and-future as found in "the legends of my people," death, beyond-death, totality. Mythologies are the original totality views of the world for each tribe, and the more isolated the tribe, the more perfectly do its myths perform the job of making order out of chaos and maintaining the tribe's sense of continuity and solidarity. For an individual to attain the highest degree of self-transcendence *within* the tribe's totality view would of course mean to serve the *tribe's* hopes or fears, benefit expectancies or harm expectancies, to the point of self-sacrifice. Short of that, however, there are many other "good things" the individual can do for the tribe, "bad things" for the tribe's enemies, in the traditional tribal roles. All such roles demand an internal ordering of the emotion repertoire on a means-to-end basis. Thus originates the *ego* as a continuous, consistent, internally or externally regulated hierarchy of values, giving priority to some emotions and repression or redirection to others. We all know how tribalism brings out the best in people and how it arrests self-transcendence at the limits of the tribe's welfare. But once the consciousness of totality is there, no matter how constricted or corrupted the form, *the die is cast*. There can be no other goal for self-transcendence than totality. Even egomania is a blinded, stupor-filled, self-arrested form of the yearning for totality. True self-transcendence works in the direction of knowing, loving, and serving the totality as one once knew, loved, and served, the family, the tribe, the nation, the ideology, the particular religion, and so on. But we are getting ahead of our story—let us return to our grade-school child and to the modern world.

One criterion of an existing value system's human adequacy would have to be the value placed on the education of the young regardless of privilege. In addition, we place before the young, in late childhood and early adolescence, a great variety of carefully considered self-ideals in the form of heroes or leaders or honored persons, *other than the parents* (for we are always dealing with a new "throw" of the genes), to help the youth to discover what he admires, to fire his imagination and energies into specific channels, and to concentrate his actual (inherited) potentialities. We see this activity take place, regardless of whether it is guided by adults, in the games, the play-acting, the pretending-to-be, the daydreaming, the hero-worshipping, of this period of life, whose importance as a trial run for later life can hardly be overestimated. Puppies and kittens play, but not like that! "In dreams begin responsibilities," and so we see that another criterion of a value system's human adequacy is the degree to which it keeps this exploratory period open, as against hustling the young prematurely into fixed social roles.

Then, however, a time comes when the youthful person must realize that, even given many possibilities, in games and dreams and make-believe he can be many different selves at once; but as a person, an individual with a biography, he *becomes* only one: the one who, in succession, first pretended to be, then tried to be, and then outgrew, these many different selves. Looking backward over his early life, a man mourns the many selves who had to die so that he might get beyond them, and at the same time he knows the continuity of his identity through them all.

A time comes—for some sooner, for others later—when the individual first begins to feel the existential squeeze-play: so much to be done, so much not to miss, and so little time to do it! It may be the experience of sickness, the loss of vitality, and the death of friends and relatives that forces some people to foresee their own end approaching; others perform all these experiences in imagination, or with the assistance of literature. The growing person now tends to become more "realistic" about himself: he sees that of all those marvelous possibilities that entice and invite enactment, some lie outside his talents (thank goodness!); some would take years of preparation, more than he can spare; for some the opportunity will not be given; for some it is already too late. The honeymoon with the dazzling possibilities is over, and it is time to settle down with the realizable goals, though perhaps haunted by the pathos of unlived lives. Unfortunately, it is only the precariousness and fleetingness of life that finally force even the least philosophical of men to think occasionally of its total meaning and even to entertain the idea of salvation in a negative form: how not to waste it. And it is precisely at this point that the greatest danger for the individual may arise. Having decided prudently not to waste his life in dreaming, and to settle down "realistically" into some social role for which he is best fitted by aptitude and education, he believes he has reached maturity and life can demand no more of him than to do his job well: "my station and its duties." We have only to remember Kierkegaard's satirical descriptions of the Christian in Christendom as a husband and father, alderman, pillar of society, advisor to the young, and so on, to get the point, which would hold true of any "official" religion. Still another criterion, then, of a value system's human adequacy is the extent to which it encourages thinking about the ultimate meaning of life beyond the fulfillment of social roles; such thinking best takes place within a framework of religious and philosophical freedom.

Through self-transcendence we approach the limits of knowledge of the human type, indicated in Figure 1*a* by the broken line. These limits are different for different persons in different historical times, and for each person in different parts of his life, and are unspecifiable either for oneself or for others, since no one knows how much more anyone will be given to live through, nor with what limitations. Although self-transcendence is sometimes described by psychologists as the process of finding and adopting a valid self-ideal, and by religious thinkers as spiritual growth, both process and growth seem quite inadequately descriptive terms for something so mysterious, so unpredictable, so dependent on apparently fortuitous "happenings," such as people one happens to run into, historical events that cross one's path, and the many unplanned, unasked-for situations in which one "finds" oneself. It is more as if the original "interaction" between the self and the world (or, more biologically expressed, between the nervous system and the nonnervous system in which it must operate) gradually turned itself into a dialogue, in which the individual *listens, responds*, and then *listens again*. Self-arrest could then be thought of as a kind of cutting-off of the conversation, stopping up one's ears, whether through fatigue or dismay, fear of hearing the worst, or having "had enough." Self-transcendence and self-arrest can no more be finally defined than the mystery on the edges of which they stand, and they involve both the knowledge-enhancing and the knowledge-hindering functioning of imagination. It is certainly the most *costly* "process" known, this human form of growing up, accumulating as it does in the past of the individual as well as in the history of the race the unforeseeable by-products and long-term results of actions taken long ago, when the agent did not "know better" or, even knowing better, did not succeed in doing better. We have only to speculate on whether the man-created misery in the world is due more to the lack of imagination (in the form of insufficient projective imagining and empathic imagining regarding the farther reaches of man's own actions) or to vain imaginations (in the form of false self-ideals or egomanias) to become convinced that man is the animal that is most in need of help. And yet without the renewed trial and error in the adoption of ideals, how would the "real" self be recognized? And how could there be a *therapeutic* disillusionment (with or without professional help) rather than *cynical* disillusionment, the most common outcome of the pursuit of ideals?

Although self-transcendence seems to be future-oriented, it is influenced by the past and yet in part it *creates* the past, as the arrows in Figure 1*a* indicate. And part of the past it creates is so harmful (i.e., enslaving, destructive, and self-arresting) that the whole "process" would quickly paralyze itself, unless there were some way to deal with the past beneficially. In the terms of our generalized formula for animal knowledge, the way to deal beneficially with the past is to learn what it is trying to teach us. But we see now that for the expanded human type of consciousness this cannot be merely to survive, as idealists have always properly, if somewhat hastily and recklessly, pointed out. It must be something more like trying to realize the highest possibility that demands to be realized in every limited situation—which is also an operational way of stating what people actually do when they believe they are "doing the will of God." And it requires not only *learning* from the past whatever can be learned, but also *forgiving* a great deal of the past, both individual and historical, in order to be liberated from it and to keep the future open for something new. The forgiveness of sins is, in this light, not a religious luxury demanded by squeamish persons with oversensitive consciences, but a *condition of growth* for this species, as any parent soon discovers. The Psalmist, in calling attention to the forgiving qualities of the Lord, merely expressed his belief that God as the ideal loving father knows this already and is trying to teach it to man.

The foregoing will have to do as our attempt to answer questions 8 and 9 affirmatively. Yes, there is a specifically human nature, from the biblical standpoint; and the metacriteria derived from it not only define beneficial/harmful for it, as against other animal natures, but supply a measuring rod to determine the human adequacy of other existing value systems. The metacriteria would not lead to that logical bogey-man, the infinite regress, because they are not the logical attributes of something, which must then be reapplied to the attributing of the metacriteria themselves (a process very few people can take more than three steps back). Instead, the metacriteria stem from man's being this kind of creature rather than another kind. Thus the proper way to disagree with them is to say: you are mistaken; man is not that kind of creature, but another kind.

The coercive power of society to punish and reward types of behavior is the immediate authority for value systems, but the ultimate authority can be nothing less than the Author of this kind of creature (question 10). Here we are not yet ready to talk in such quasi-theological or cryptometaphysical terms.

Question 6, about the language of moral discourse—be it descriptive, imperative, prescriptive, commendatory, ends-and-means-relating, goal-defining, conduct-guiding, or society-regulating—surely depends on how broadly that subject is to be treated and what specific parts are to be talked about. In the broad sense of our treatment of the subject of values as the meaning and function of the animal value system extended to the human case, the language has been mostly descriptive; but the description has pertained to emotions and actions regarded as evaluations of situations so complex that even the functions of language just mentioned may be exhausted in trying to reach verbal understanding. If knowledge of values is the mapping of actions-in-a-situation as beneficial or harmful, the language for it must be intentionally action-guiding no matter how various in grammatical form, and also attitude-guiding insofar as attitudes are dispositions toward action.

A preliminary step toward avoidance of verbal muddles would be to translate ethical or moral recommendations containing "good" and "evil" from the thing-language of values to the action-language of evaluation; and I leave this as something for the reader to try out, sometimes using merely harmful/beneficial, other times the expanded criteria. (Example: Do not be overcome by evil, but overcome evil with good, would translate as follows: do not let yourself be defeated by actions and attitudes, your own or others', that are enslaving, destructive, and self-arresting; rather, overcome such actions and attitudes by actions and attitudes that are redemptive, preservative, and self-transcending.)

It goes almost without saying that Christians are, at the least, people who see in the life and teachings of Jesus of Nazareth the historical "showing" of just the kind of action-and-attitude guidance that man needs; and therefore they regard Jesus as sent from God, Who, in His transcendence, is for such guidance too hidden, too Other, too wrapped in the impenetrable mystery that is the permanent medium in which the human consciousness flickers.

But entirely aside from religious interpretations, the overwhelming fact that we are faced with in the species-specific human nature is that in it the animal value system that underlies evolutionary changes has achieved consciousness: self-consciousness; others-consciousness; past-, present-, and future-consciousness. Thus are revealed certain *inhuman* aspects of nonhuman animal value systems, making modifications necessary just to keep human consciousness bearable. The animals do not know what they have to do to other sentient creatures to survive, and they have not the imagination either to be cruel or to do unto others as they would be done by. They are exonerated. But man knows what he must do to survive, and he knows it more fully, as his consciousness expands in the direction of totality. We now turn to the problem of totality to see how it has been and can be dealt with cognitively.

Seven The Unique Problem of Totality: Is an Ontology Possible?

I

History, as Kierkegaard observed in one of his tirades against Hegel, has a way of immortalizing its villains as thoroughly as its heroes, and unfortunately for us, history has also succeeded in immortalizing certain terminological confusions originating in the most trivial literary accidents—such as the confusion between ontology and metaphysics which is our philosophical heritage. As any college boy knows, the editor of Aristotle's collected works named the sections coming after those dealing with "nature" or "physics" *meta ta physika* (after the physics), and as *metaphysics* they have been known ever since, although their subject matter was called by Aristotle himself "the science of first principles" or "the science of being *qua* being." He even expressed some quite modern-sounding doubts and name-calling with respect to the possibility of such a science:

> One might be tempted to say that the acquisition of this science of first causes is beyond the power of human nature, which is servile in many respects. In that case, as Simonides says: "God alone can

have this privilege," and man should only seek knowledge proportionate to his limitations. If there is anything in what the poets say, and God is by nature jealous, He must be particularly jealous in this matter of wisdom, and woe to those whose knowledge is above average! Actually, however, the Godhead is incapable of jealousy, and, as the proverb says: "Poets are not seldom liars." (1)

So, historically, Aristotle was made to "do" *ontology*, the *logos* of *to on*, under the title of *metaphysics*. His predecessors did not call their speculations about the general nature of the most fundamental reality either ontology or metaphysics, but rather the theories, the doctrines or the teachings of this or that *school*—as, say, the school of Pythagoras, the schools of Protagoras, Parmenides, Heraclitus, or Plato. In particular Plato, whose mantle of unsolved problems Aristotle inherited, did not call his speculations on the nature of *"ta onta"* either ontology or metaphysics, but rather the Doctrine of Forms or Ideas. The Doctrine of Forms or Ideas was a tightly interwoven combination of what we would now call cosmology, epistemology, and logic, and Plato expected that it would give a single, unified answer to at least four distinct questions: why the world is the way it is; how man can get from mere opinions on appearances to true knowledge of the "really real"; how the definitions of ordinary words can be syllogistically clarified to yield their kernel of eternal, unchanging meanings; and how universal properties are related to the particulars on which they are predicated. Whether this ambitious effort would nowadays be called an attempt at ontology or metaphysics depends on how we interpret the "meta" in metaphysics. After Aristotle's "metaphysics" carved out the province of the science of first principles as a concern with being *qua* being, the "meta" in metaphysics came more and more to mean, not "after", but *beyond*, and *beyond* in several different senses: beyond as *transcending*, beyond as *logically prior to*, beyond as *cosmologically prior to*, beyond as *more funda-*

mental than, beyond as *more inclusive than*, the "natural" world studied by natural science, or even the world of ordinary human experience. With such a broadening of meaning thanks to the several connotations of "meta" as beyond, the original misnomer, metaphysics, became the candidate selected by history for the wider role, slated for proliferation and specialization into several departments, namely, cosmology, ontology, theology, logic, epistemology, and nowadays, philosophical anthropology and phenomenology.

To make matters worse, both Plato and Aristotle pursued an undisguised theological interest in their ontological (metaphysical) speculations: Plato by making his Forms the divine paradigms which the Craftsman uses to create the actual world from resisting matter, Aristotle by asserting that the study of the highest genus, being, *necessarily* becomes theology when the study of its first principles, or causes, reveals the Prime Mover, God, who both sets everything in motion (efficient cause) and, as the highest good, provides its goal (final cause). These historical and inescapable involvements of ontology and metaphysics in theology, not to mention their subsequent adventures in would-be Christian systems of theology—ontology more in the Platonic or Augustinian tradition, metaphysics more in the Aristotelian or Thomistic tradition—have given the two words a guilt-by-association odor of sanctity that is obnoxious to the free and uncommitted temperament of modern philosophy, even where it is not rabidly antimetaphysical. Even a well-received theologian like Paul Tillich is often thought to be saying something important in spite of, rather than because of, his ontological vocabulary. Add to this the literary and academic Ordinary Language habit of using the adjective "ontological" to mean anything the least bit profound or deep-searching, and the adjective "metaphysical" to mean anything the least bit abstruse or impractical, and it is evident that the linguistic fog we are starting out in is a real pea-souper. Not surprisingly, those who *are* rabidly antimetaphysical simply equate ontology and metaphysics, on the cogent grounds that things equal to the same thing—namely, nonsense—are equal to each other. Among these, the more tolerant admit that, man being the linguistic invalid that he is, a certain amount of nonsense is bound to get babbled, and some of it is bound to be profound nonsense (ontological) and some abstruse nonsense (metaphysical). In this chapter we generally prefer to use the word ontology, precisely because it is narrower and the case against it is more severe, there being less escape into vagueness by means of the "meta" connotations of metaphysics.

Our task is not to construct an ontology, but to see if an ontology is possible at all in the cognitive sense, and to use those features of the present epistemology that might warn us of at least the worst of the errors and pitfalls of past ontologies, such as any future ontologists would wish to try to avoid. Thus in the first sections of this chapter we write a kind of Greek prologue in the form of a cautionary tale. We honor the pioneers, not by eulogizing them, but by searching out what is correctable in their "early maps," leaving to history what can no longer be used, and allowing ourselves to be instructed by the very nature of their errors to see if we are not unwittingly repeating them, under different auspices.

Any reader who has followed the constructive part of this epistemology with some conviction and sympathy ought now to be able to go back and reread his Plato as if the scales had fallen from his eyes, at least as far as the Doctrine of Forms is concerned. For he should now be able to see what Plato was *doing,* as neither Plato nor his disciples could possibly have seen. To the modern mind the Doctrine of Forms, baldly stated, sounds like a fantasy-flight of exalted nonsense, coming from an otherwise sober, practical, sense-talking man: it is the belief "in eternal, unchanging, universal absolutes, independent of phenomena; in, for example, absolute beauty, absolute justice, absolute goodness, from which whatever we call beautiful, just, or good derives any reality it may have." (2) If, however, we temporarily erase from our minds all those quasi-scientific ideas which have become part of our common-sense world, and by imaging-as-if-perceiving and plenty of empathic imagining project ourselves back into Plato's world, we see Socrates and his contemporaries asking themselves questions about their common-sense world very like those asked by intelligent, curious children today, as soon as their language has provided them with a referential schema or map of the *furniture* of their world: What are things? What are people? What is everything made of? What is good and bad? Why do things change? Does everything change, or is there something unchangeable that only seems to change? What makes some things similar, other things different? Our children, however, do not get very far with their questioning, for we rush to them a scaled-down, simplified scientific answer to each question as it comes along, thereby not only stifling their natural sense of wonder (every question obviously has some scientific answer!) but preventing them from becoming what the existentialists claim that man is supposed to be: the being that asks the question about his own being. Or at least we put off the time for a while.

But in an age when every man had to be his own scientist, or the disciple of one of the many conflicting schools, he got to the ontological questions much sooner.

Now Socrates noticed that the answers given in his day were all in the nature of opinions, *doxa*; that is, relatively good or bad guesses, such as everything is made of water, or of numbers, or of atoms, or of elements like earth, air, fire and water; everything attracts or repels everything else; everything changes, or, again, nothing changes. Therefore Socrates asked, instead, what sorts of things *would* the world have to be made of, in order that *knowledge*, rather than mere guesses, about those things would be possible? The question was posed more specifically as an attempt to find a refutation of Heracliteanism, the doctrine that everything is in a state of flux, activity, process, as far as it is reported by our senses. Now this theory would certainly explain why there are all these conflicting opinions, but if carried to the extreme (conflicting with what?) it would make knowledge impossible, and even good guesses meaningless. Even the most elementary description of physical happenings would be impossible if all the parts of the happenings changed during the description, and the words of the description also changed, and even the describer himself changed.

I. M. Crombie calls this extremist view "rampant Heracliteanism" as against "normal Heracliteanism" which merely states (trivially, he thinks) that some things in the world are more stable, some are less stable. Crombie suggests that Socrates ought not to have been influenced by this extremist view to the extent of denying sense perception the status of knowledge (see *Plato's Doctrines*, Vol. II, pp. 10–12, *The Discussion of Heraclitus*). He points out that

> [The blunder of the rampant Heracliteans] (if there were any) is that of confusing "all properties result from change" with: "all properties are subject to change." Perhaps an illustration would help. An electric bulb glows (we will suppose) because of some kind of incessant activity in the filament. But although the glowing is a process which *results from* activity or change it is not in itself a process *of* change, in the way in which a continuous flickering could be said to be a process of change. The Heraclitean doctrine which Plato is refuting amounts to the doctrine that, since the incandescence of the bulb is due to activity in the filament, there can never really be a steady glow but only a flickering one. (3)

What Socrates could not possibly have known (and perhaps Crombie did not either) is that it is precisely the unaided senses, operating on the everyday level, that produce the gestalt effect of a steady glow, and also the many other persistent, fused gestalts of sensory qualities we call *stable objects*. Apparently this understandable ignorance led Plato to adopt a sense-datum theory to the effect that some sort of nonsensory *conscious mental process* is needed to convert the sensory flux into enduring things, which alone can be the objects of knowledge. As Crombie puts it:

> [This *aisthesis*,] this perceiving of sense-data is common to all organisms, and it does not constitute knowledge of the external world. Knowledge of the external world only arises when we notice the occurrence of our sense-data and, by comparison of one with another, assess their significance as pointers to our future experience, when we notice for example that we are perceiving kinds of objects to which we have learnt to give the name "black clouds", and conclude from this that we shall shortly experience what we have learnt to call rain. (4)

We are tempted to ask, well then, how do the animals get around in the rampant Heraclitean flux? We don't see them crashing into trees, or unable to assess the future significance of the sight of food! But that's not fair, if we are really projecting ourselves back into Plato's world.

Let us admit then that rampant Heracliteanism (which is self-refuting, since it can certainly be described) should not have made Socrates so "nervous," as a threat to knowledge of any kind. But if we now think back to Chapter Two where we described the barrage of discriminable differences fed into the human nervous system by the sense of sight and extended to limitless proportions in imaginative contemplation, we see what Plato was up against: rampant Heracliteanism was his vision of chaos. We said in Chapter Two:

> The discriminable differences simply will not be denied. The eye, that is, cannot deny what it itself discerns or perceives; and this burden of particulars, this overwhelming complexity, is passed on to the mind with a mandate for the mind to *find* the order in it. And the mind naturally turns toward knowledge, expecting to find through knowledge the order in or behind the chaos, for that is one of the defining characteristics of what man thinks knowledge to be: knowledge is that which makes order out of chaos.

With the best possible epistemic motive, Plato set out to find something fixed and unchanging in which to trap and contain the chaotic flux of sensation and produce something worthy to be called knowledge; thus he unwittingly became the Columbus of the Categories. And like Columbus, he did not know what he had discovered, for it was not categories, the Forms, that he actually discovered, but the *categorizing process in the brain*, interacting with the categorizable world. This process in the mind had already deposited its *products* in certain parts of the language, especially in the abstract nouns standing for the abstractable properties of things and events: beauty, justice, triangularity, equality, oddness, redness, and so on, *ad infinitum*. Furthermore, in order to make sure that the defense against Chaos would hold, Plato had to build these "forms" into the ultimate structure of the universe by means of his cosmology. In their purified (perfect and absolutized, eternal, unchanging, universal) versions, which it was the goal of dialectics to achieve, these *abstract definitions* became the models or archetypes which the Craftsman looked at while constructing the visible world out of resisting formlessness, the degree of imperfection in this world being thus explained by the degree of resistance put up by matter.

A sound intuition told Plato that these forms, *ousiai*, or "intelligible natures" that made knowledge possible, could not be learned during the individual's lifetime, for what could they be learned from, except the fluctuating reports of the senses? So, knowing nothing of heredity or evolution, he invented the Myth of Recollection: the Forms were remembered by our souls from a prenatal journey around the Platonic Heaven where the Forms existed eternally in the direct sight of the gods; and after our birth, the Forms were recollected when their imperfect images here reminded us of them. Crombie whimsically calls this part of the theory "pre-natal geometry lessons," but there is some truth in it. We now know, for example, that the child does not *learn*, but only *puts to work* the processes in his mind that make geometry lessons possible. Indeed, it is these same processes of categorization, such as identity-categorizing or "forms of the same thing" and equivalence-categorizing, or "the same kinds of things" which convert, for the animal as well as for the preverbal child, what otherwise would be a "rampant Heraclitean world" into a "normal Heraclitean world." If this were not the case, we could not teach language to a child, for there would be no nonverbal category to point to, to act out, or to pantomine, to correspond (roughly) to the speech category.

II

But now we must observe how this excellent pioneering start in epistemology became changed into something quite different under the pressure for order that comes from the fear of chaos, the temptation to order-at-all-costs. It will probably never be possible to decide to what extent Plato himself regarded the theory of Forms as an epistemology, to what extent as an ontology; but there is no doubt that it was as an ontology that this theory played its most important role in his theology, and in all those theologies in the Platonic tradition, Jewish, Christian, pagan, and perhaps even Islamic. The theological problem was: how to make order out of the order-making forms themselves; how to arrange them in an order of *value*, corresponding to some recognizable order of *structures*; and if that is what you think the ultimate order must be, you could hardly do better than Plato did. Let the Form of the Good be the highest form, at once the order-making form which creates the other forms and the form in which all the other forms participate in so far as they are good. Now we not only have an abstract definition of God, the unity of the highest being with the highest value, but surely we have something that deserves to be called absolute, eternal, unchanging, universal, independent of phenomena, and the source of whatever reality the other forms may possess. What is more, we have an "intelligible nature" or *ousia* of God that anyone can understand if his brain is in working condition. As the sun is in relation to the objects of the physical world, so the Form of the Good is in relation to the objects of reason, for by the light of this source we both see the other objects, and we see what illuminates them so that we can see them. The Myth of the Cave tells us that in this world, we (most men) cannot look at the sun directly; but it tells also in what direction we must look to see the sun better, rather than mere shadows. In NeoPlatonism the Sun became both the One and the Good, from which all else "emanates" or "radiates."

It should be scarcely surprising to us that the man who expected abstractions to make possible the only reliable knowledge, and by the same token, to create order out of chaos (or to *be* the order behind the chaos), should eventually fall in love with abstractions, and that even his idea of love (Eros) should bear a strong resemblance to the process of abstraction itself, as described in Chapter Five. Abstraction is the process of *subtracting* from particular objects or events some feature that can be characterized independently of them and then, with the aid of projective imagining, setting this feature up as the criterial attribute of a wider class or category (including many otherwise differing members) the advantage being precisely the generality or universality that is built into them at the expense of particularity and concreteness. Here is what Diotima tells Socrates concerning the greater and hidden mysteries of Eros in relation to its desired object, Beauty:

> "It is necessary," she said, "that one who approaches in the right way should begin this business young, and approach beautiful bodies. First, if his leader leads aright, he should love one body and there beget beautiful speech; then he should take notice that the beauty in one body is akin to the beauty in another body, and if we must pursue beauty in essence, it is great folly not to believe that the beauty in all such bodies is one and the same. When he has learnt this, he must become the lover of all beautiful bodies, and relax the intense passion for one, thinking lightly of it and believing it to be a

small thing. Next he must believe beauty in souls to be more precious than beauty in the body; so that if anyone is decent in soul, even if it has little bloom, it should be enough for him to love and care for, and to beget and seek such talks as will make young people better; that he may moreover be compelled to contemplate the beauty in our pursuits and customs, and to see that all beauty is of one and the same kin, and that so he may believe that bodily beauty is a small thing. Next, he must be led from practice to knowledge, that he may see again the beauty in different kinds of knowledge, and, directing his gaze from now on towards beauty as a whole, he may no longer dwell upon one, like a servant, content with the beauty of one boy or one human being or one pursuit, and so be slavish and petty; but he should turn to the great ocean of beauty, and in contemplation of it give birth to many beautiful and magnificent speeches and thoughts in the abundance of philosophy, until being strengthened and grown therein he may catch sight of some one knowledge, the one science of this beauty now to be described. Try to attend," she said, "as carefully as you can."

"Whoever shall be guided so far towards the mysteries of love, by contemplating beautiful things rightly in due order, is approaching the last grade. Suddenly he will behold a beauty marvellous in its nature, that very Beauty, Socrates, for the sake of which all the earlier hardships had been borne: in the first place, everlasting, and never being born nor perishing, neither increasing nor diminishing; secondly, not beautiful here and ugly there, not beautiful now and ugly then, not beautiful in one direction and ugly in another direction, not beautiful in one place and ugly in another place. Again, this beauty will not show itself to him like a face or hands or any bodily thing at all, nor as a discourse or a science, not indeed as residing in anything, as in a living creature or in earth or heaven or anything else, but being by itself with itself always in simplicity; while all the beautiful things elsewhere partake of this beauty in such manner, that when *they* are born and perish *it* becomes neither less nor more and nothing at all happens to it; so that when anyone by right boy-loving goes up from these beautiful things to that beauty, and begins to catch sight of it, he would almost touch the perfect secret. For let me tell you, the right way to approach the things of love, or to be led there by another, is this: beginning from these beautiful things, to mount for that beauty's sake ever upwards, as by a flight of steps, from one to two, and from two to all beautiful bodies, and from beautiful bodies to beautiful pursuits and practices, and from practices to beautiful learnings, so that from learnings he may come at last to that perfect learning which is the learning solely of that beauty itself, and may know at last that which is the perfection of beauty! . . . " "What indeed," she said, "would we think, if it were given to one of us to see beauty undefiled, pure, unmixed, not adulterated with human flesh and colours and much other mortal rubbish, and if he could behold beauty in perfect simplicity? . . . !" "Do you not reflect," said she, "that there only it will be possible for him, when he sees the beautiful with the mind, which alone can see it, to give birth not to likenesses of virtue, since he touches no likeness, but to realities, since he touches reality; and when he has given birth to real virtue and brought it up, will it not be granted him to be the friend of God, and immortal if any man ever is?" (5)

It appears from this spelling-out of the process of abstraction that our Columbus of the Categories was misled—both by his search for something eternal and unchanging with which to battle Heracliteanism and by his understanding of mathematical abstractions as these were expounded in his day—into assuming that *all* abstractions, provided they were sufficiently clarified and purified by the dialectical process, would result in the kinds of abstract definitions that are to be found in mathematics. If the reader will perform the experiment of running through this description again, substituting everywhere triangular bodies for beautiful bodies, and triangularity for beauty (temporarily ignoring the confusion of categories that starts at "next he must believe beauty in souls, . . ."), he will see that the whole passage makes a great deal of sense even today, especially for an emotionally exalted mathematician, who perhaps in moments of "enthusiasm" believes that God is a mathematician. For surely at "Suddenly he will behold . . ." he will arrive at the Science of Triangularity, which is not only independent of triangles in space and time but also contains all those internal relations, such as the Pythagorean and other theorems about triangles, and all of trigonometry, that we would never suspect just by looking at triangular bodies. Nowadays we tend to regard properties like oddness and evenness as belonging to numbers, which themselves are abstractions from countable groups. Therefore the properties named are abstractions of abstractions and constitute a set of categories that can in turn be manipulated by abstractions of *operations,* such as adding and subtracting. In this way we rapidly approach a world of partly empirical, partly invented abstractions, so far removed from the actual world that it can well afford to be independent of it. It is not necessarily a world of which the empirical world is an image or resemblance. Only when we force a *suspected* resemblance, when we say *let* the radius of the earth be the r into $2\pi r$, do we get any knowledge of the Heraclitean world out of mathematics and can we properly speak of a "mathematical description" of the world.

The moral of this first and quite understandable error of Plato's, and the *caveat* for future ontologists, is not that there is something wrong with abstractions, but that there are many degrees and levels and kinds and interdependencies of abstractions, depending on the "field" or area of interest in which they have been developed or have been found rewarding or useful. In our discussion of the categorizing process in Chapter Five, we found that the classifications we talked about themselves had attributes: for example, we discussed identity categories, equivalence categories, affective, functional, and formal categories. However, the attributes of the categories were not the attributes of the members of the categories; rather, they described the areas and purposes for which the categories were formed. This brings us to the second mistake Plato made: because he thought he was discovering some *things*, to be sure nonsensory things—the forms or models according to which the actual world was made, rather than a process in the mind productive of classes—he thought that the criterial attributes of the members of the class were attributable to the class itself. This is the reason for all the talk about chairhood-itself, bedhood-itself, triangularity-itself, also beauty-itself, and justice-itself. Now, although it is a speech-habit sanctioned by Ordinary Language usage to say simply "chair" as a shorthand for "a member of the subclass of household furniture whose criterial attribute is to be used to sit on," nobody would claim, on careful reflection, that the thing named characterizes the class itself. The *class* of chairs is not a chair, the class of triangular things is not triangular, the class of beautiful things is not beautiful, any more than is the class of evil things evil, the class of dirty things dirty, or the class of malodorous things bad-smelling. It is like thinking that the manila folder holding in a filing cabinet unpaid bills had to be an unpaid bill "itself" or the folder holding the income tax reports had to be an income tax report.

This leads us right into the classical (medieval) problem of Logical Realism versus Nominalism, a question we promised in Chapter Five to discuss here, because the present epistemology appears to be so utterly against realism and on the side of nominalism. The Logical Realists claim that universals—that is, the names of genuses, species, and properties, (man, chair, red, etc.)—are objectively real constituents of the objective world, independent of particular examples and of particular minds knowing them, whereas the nominalists claim that universals are "nothing but names" and that only particulars exist for our minds. Obviously the Platonic Doctrine of Forms is the origin of logical realism, and by now this controversy should be obsolete, but not because the nominalists have won. Nobody has won. Classes are just as real as the items being classified, but they are realities of a different kind. The catalogue cards in a library are just as real as the books that they classify, but they are realities of a different kind, and to say that the particular books are real but the cards showing their subdivision and arrangement in the library are "nothing but names" is to slight the function of names. Isn't that the very least that names should do? On a more physiological level, there are no doubt neuronal patterns in the brain that correspond to the classes, neuronal patterns that correspond to things being classified, and neuronal activities that correspond to the process of classifying whatever is classifiable, in whatever way seems desirable. Thus the only remnants of this controversy that still make sense are the questions of the degree of permanence or change and common consent on classes and items classified. In the case of words, it is the need for communication that reduces the amount of idiosyncratic classification allowed before they stop functioning as message carriers. Communal interaction enacted in dialogue, not logical realism, keeps us from going privately insane.

The third mistake our Columbus of the Categories made was an indirect result of the first mistake. If we do not distinguish between different degrees of abstraction and different kinds of materials abstracted from, we are apt not to notice that the loss of particularity that is involved in the very process of abstraction, which is the price paid for the generality, matters very little in the case of abstractions like triangularity, which is not in any great danger of being confused with squareness. Loss of particularity, however, matters a great deal in the case of abstractions derived from human feelings and behavior and from social realities. This danger becomes painfully evident when we see how Plato has confused the categories starting at "Next he must believe that beauty in souls is more precious than beauty in the body. . . ." Here he confuses beauty first with virtue, then with justice, and then with truth, in an order of ascent that obviously represents his "scale of values." Assuming that an important criterial attribute of the attraction that the lover feels toward the beautiful body is, as Plato himself says, that he finds it aesthetically "entrancing," what is so aesthetically entrancing about a man's being honest instead of dishonest? What is so aesthetically entrancing about knowing that a crime is committed and the criminal is apprehended, judged, sentenced, and executed under a just law of the state? What is so aesthetically entrancing about the fact that knowledge tells us, among other things, that a certain disease is incurable, and the loss of a friend unavoidable? Did Plato find the trial and execution of Socrates aesthetically entrancing? Ennobling, perhaps; saddening, surely, and awe-inspiring—but not beautiful! However, at once we must remember that the confusion of categories that could easily become disastrous in the world of happenings, actions, and responses, matters not the least in the world of Forms to which Plato is leading us, where nothing happens, nothing changes, nothing is endangered. Plato the political moralist thought that men would become more just if they could "see with the mind" the eternal Form of Justice (as indeed they would become better mathematicians if they could understand Triangularity)—but for this to make sense, we have to hold firmly to his theological model of the Myth of the Craftsman and the Myth of Recollection. In the Platonic heaven of the prenatal journey there is no need for the forms to be either clearly distinguished or interrelated, for they are all forms of the Good, the supreme order-creating value. (It will be evident to the reader that Plato also committed the usual reification of the concept of value; but since this is certainly not peculiar to him, but more like a general affliction of mankind, we treat it as a problem for the ontologist later.)

The reader will also have noticed by now that the passage by "stepping stones" from earth to heaven described here is at least one version of the ladder of ascent described in various mystical religions as the means by which the individual soul seeks to achieve union with the Godhead, or the atman with brahman in the Nirvana of Buddhism. However, it would be wrong to regard Plato as a "practicing mystic" in the sense later developed from his philosophy by Plotinus, and, still later, the medieval Christian adaptations of Plotinus. Apparently Plato had moments when he was caught up in an emotional exaltation, described as a "divine madness" in the *Phaedrus*, and the passage we are studying represents the culmination of one such ecstatic approach to the divine. The Eros described by Diotima earlier as a hunger or yearning for that which one is lacking, and as a disguised form of the yearning for immortality, is a mythologized form of the hunger for abstraction, of a yearning to pass from the particular to the general, from the concrete to the abstract, *because in that direction lies the universal, and the universal is identified with the divine*. Identifying the universal with the divine is a mistake made by virtually all idealistic philosophies, when, as we have seen, the categories that are the most easily universalizable are likely to be those which are abstracted from the most simple, humanly unimportant, and even trivial aspects of the experienced world. We must credit Plato's mesmerization by abstractions and their supposed nearness to the divine to a kind of early narcissism of the mind discovering and falling in love with its own powers. Later we learn what happens when we make this mistake today.

When we turn to Aristotle as the second pioneer to see what misled *him*, it is obvious that he was as yet in no position to see clearly what had misled Plato. However, after long hesitation (20 years), he decided that the Theory of Forms would have to go, mostly because the Forms did not *explain* anything in the actual, sensible world, except its imperfection. At its worst, the theory simply *doubled* the number of things that had to be explained, the imitation on earth and the self-subsistent paradigm in heaven. Aristotle saw the mistake Plato had made about mathematics, but not the way we see it:

> How are we to be convinced that the numbers have separate existence? According to the school of Plato, they act as causes of existing things, since each number is an Idea [Form] and the Idea [Form] is somehow or other related to other things as cause to

effects—so at least the Platonic theory presupposes . . . What causal value has *this*? [Number] is not *asserted* to be the cause of anything, but a self-subsistent entity: nor is it in fact *observed* to be the cause of anything. The theorems of arithmetic, as I have said, are true of sensible things and do not imply self-subsistent mathematical number. (6)

Here we have Aristotle's problem in a nutshell: how to get those of Plato's Forms which seemed to have some explanatory value other than imperfection down from that transcendent heaven and firmly installed or imbedded, or made immanent, in the sensible world? Most difficult of all would be to get explanatory ideas of *motion* and *change* out of Plato's Forms, when the entire theory had been devised to stop or deny the Heraclitean flux, which Aristotle thought stood in need of explanation, not denial. On the other hand, when it came to theology, Aristotle could see the advantages of Plato's "eternal, unchanging, perfect," as applied to those Forms that were considered to be aspects of the divine. Aristotle could have helped himself out by resorting to the Myth of Recollection for the divine attributes only, but he was too good a psychologist to invest human memory with any such divine-remembering task; and therefore he also gave up the myth of the prenatal journey and the Form-guided Craftsman-creator. No, the divine would somehow have to be fitted into the same causal scheme that would explain the sensible world, but in some extrapolated or exalted version. Aristotle's solution was to apply the four kinds of causes that he had already "dealt with" in his *Physics*, to the highest category, *being*, in order to produce what he called *first* causes and thereby the science of being *qua* being:

So far I have explained the nature of the science of first causes, as well as its object towards which all our studies no less than our present inquiry are directed. Clearly we must endeavor to acquire this knowledge; for once we can claim to know the primary cause of any particular thing we know that thing itself. Now we recognize four kinds of cause: (i) the definition, essence, or essential nature of the thing; (ii) its matter or substratum; (iii) its source of motion; and (iv) opposite to the third, the "end" or "good," which is the goal of all generation and movement. (7)

Regarding (i) Aristotle also says: "The definition is the final thing to which the 'reason why' is reducible, and the ultimate reason why is a cause or principle [and therefore the definition or essence is a cause]." Aristotle does not want us to think that the essence is a paradigm case in heaven as in Plato, yet this is the "trace" left on his thought by bringing the Theory of Forms down into the sensible world. These are the four famous Causes of Aristotle, called the Essential, the Material, the Efficient, and the Final, later adopted by St. Thomas and used to "prove" the existence of God. A closer look reveals them to be, not so much a science of anything, but a *model of explanation*: if you can find candidates for these four causes for any particular thing, you have explained it. If you apply this to the highest category, being, you have explained it, and that is all that a science can do. Ontology is therefore *the superscience*, to which all others are subordinate as species to genus—exactly what Aristotle intended and what St. Thomas meant by calling theology "the queen of the sciences."

(It took modern science 500 years to liberate itself from this Thomistic–Aristotelian model of explanation, and it was against this "theory" of science that the positivistic movement directed its most violent attacks. So it is not likely that the modern temperament, either in those who are for or those who are against science, will be able, in the foreseeable future, to stomach any version of ontology or metaphysics that puts itself forth as a superscience.)

The chief charges against Aristotle's concept of ontology (metaphysics) regarded as *caveats* for any ontologists of the foreseeable future, are (1) substantialization, and (2) animism.

The universe consists of various kinds of substances (the formal cause plus the material cause), and these different substances are pushed and pulled around by various kinds of actions for various purposes (the efficient cause and the final cause). The whole schema is faintly reminiscent of the child's world (especially the child who is just learning to use the language), which also consists of things and their attributes and actions and their attributes and goals, encoded in most languages as roughly, the nouns, the adjectives, the verbs, the adverbs, and the valuation words (good! bad! do! don't!) Aristotle probably would have not minded this exegesis, for he regularly appealed to common sense, and the common-sense world is the expanded child-world. But as it turned out, modern science is not the refinement of the common-sense world that most common-sense people think it should be, and to the modern mind disciplined by a scientific education, Aristotle's metaphysics sounds like *animistic physics*. This means that for the physical and biological sciences it seems too *animistic*, full of mysterious push-pulls that do not explain anything, whereas for the sciences and studies of strictly human activities, it is too *physicalistic*, too substantialized, too common-sense-thing-oriented, as Heidegger objected. And it is surely in part modern physics that should be given a vote of thanks, especially from existentialists who insist on calling it meaningless, for having liberated us from the idea of a thing as a collection of plastered-on attributes held together by a metaphysical glue, the substance.

Having thus populated the universe with a large variety of things and actions, all mysteriously regarded as "causes," Aristotle put himself in a position of having to find the order in them, and the simplest way to make order out of any collection of different things is to arrange them by degrees of similarities and differences and to come up with some version of the genus–species classification. Let the reader think for a moment of all the things he encounters in everyday life that are arranged in a genus–species classification, without any claim being made for a physical or even organic relation between the items so ordered: the books in the library, the groceries at the supermarket, the furniture in the house, the merchandise in any catalogue, the goods in any "department" store, the advertisements in any newspaper, and so on. When our astronauts arrive on the newest planet they will not have the slightest difficulty classifying the strange things they find there in some version of the genus–species classification, for they will bring the equipment for this with them, in their brains. However, they will have learned to distrust this type of classification as indicating anything but the most preliminary, perhaps morphological, aspect of the strange things, and they will leave to the scientists back home the task of studying more thoroughly the wonders they discover.

Now, just as Plato had been misled by the nature of mathematical abstractions to think that he had found a world of eternal, unchanging, perfect definitions above or behind the Heraclitean flux, so Aristotle was misled by the genus–species classification to think that he had found the "explanation" of the more stable *items* in the Heraclitean flux, now regarded as a collection of moving things. The result was a pyramid, or hierarchy, of beings, from the *infimae species* at the bottom to the highest genus at the top. This is indeed how the genus–species classification of the plants and animals got started. It was not until the job was well under way that evolution as the explanation for the present distribution of higher and lower forms suggested itself, and we regard the classification as nothing permanent but merely a temporary cross section of a very slow (to our perception) process of the appearance, differentiation, and disappearance of discriminably different forms of life. How far Aristotle was influenced by his own observation of biological forms it would be hard for us to say, but certainly more than Plato. At any rate, in order not to be stuck with a static pyramid, or at most a rattling one, he needed a powerful upward pull, and for this he adopted Plato's Eros and distributed it over the entire hierarchy (Eros now being understood as the yearning for perfection). Each level of being, as matter, yearned for the perfection of its form and thereby became matter for the next higher level of being, yearning for the perfection of *its* form, and so on, all the way to the top. The top genus, perfect perfection, the Unmoved Mover or God, like a magnet attracted the entire hierarchy of beings to itself, but, being perfect, was not attracted to anything.

> The final cause, then, moves by being Loved, while all other things that move do so by being moved. . . . The unmoved mover . . . has no contingency, is not subject even to the minimal change (spatial motion in a circle) since this is what it originates. It exists, therefore, of necessity; its being is therefore good [because, being absolutely noncontingent, nothing can happen to it contrary to its nature] and it is in this way that it is a principle of motion. (8)

What does the Unmoved Mover do in its perfection? It is the Supreme Intellect, and therefore, "since the supreme intellect is the best thing in the world, it must think itself; its thinking is a thinking of thinking." Does man have any part in this? Indeed yes, for the whole physical world depends on the principle of the Unmoved Mover thinking the divine thought of thinking.

> It is a life which is *always* such as the noblest and happiest we can live [i.e., when we are engaged in metaphysical thought]—and live for so brief a while. Its very activity is pleasure—just as waking, perceiving, thinking are most pleasant because they are activities, and hopes and memories because they are hopes and memories of these. Thought which is independent of lower faculties [i.e., the divine thought] must be thought of what is best in itself; i.e., that which is thought in the fullest sense must be occupied with that which is best in the fullest sense [and therefore with the divine mind itself]. (9)

Again we arrive at a kind of exaltation of the human intellect extrapolated to the divine at the top end, but with the difference that now Plato's Form of the Good has been "immanented" into all the lower levels of being as the Final Cause, the teleology that drives each level of being to seek the perfection of its kind. Here we see Aristotle at great pains to reinstall in the structure of being what Plato had been at great pains to remove—the becoming of all things as a sort of *direction of motion* in the Heraclitean flux, which is at the same time a *graded evaluation* of the whole. Now this makes for a grandiose vista, to be sure, but it also raises even more acutely than in Plato the question of whether values can be thought of as a structure of any kind, immanent or transcendent, let alone as a structure of the genus–species kind.

Here we must leave the pioneers. We do not laugh at Columbus for having been misled by Ptolemy's map, nor even at Ptolemy's map for not showing the Americas. We might be tempted to laugh at anyone who tried to get around the world with this map today—but it would be more to the point to send out an S.O.S.!

III

It is time now to return to the mapping model of the correspondence between the two parts of reality described in Chapter Five. We want to see what sort of ontology it would allow, or legitimize, as constituting a kind of knowledge. In the terms provided by this model, the obvious analogy is a map of the whole world, the totality of the earth's surface presented with as little distortion as possible (i.e., on a globe). Totality, then, the concrete all-that-is, all-that-was, all-that-will-be, is what we now call the *referent* when the word "being" is used in the ontological sense, as by Aristotle, or when the word "cosmos" is used for the totality of the Ideas and Forms in Plato's heaven plus their earthly imitations. The mapping model provides us with a proper subject matter for ontology, namely, totality. At the same time, however, it gives us a set of procedures involved in map making which spell out the limitations, the tasks, and the impossibilities that must be recognized, if the attempt at totality mapping is not to issue in vain imaginations or in the self-glorification of the human brain, rather than in knowledge of a strictly limited sort. All knowledge is limited, as we now know, by the very nature of the receiving apparatus and the categorizing-and-mapping process, but knowledge of totality is limited in the additional sense of approaching and being concerned with a region that is the boundary line between knowledge and mystery. Paradoxically, in order not to be self-deluded, this knowledge must include acknowledgment that there is something unknowable. And here we must not make the idealistic mistake of supposing that to acknowledge a limit is to transcend it—a color-blind man might well acknowledge a limit by admitting that other people distinguish what he cannot distinguish; but this knowledge of his limits does not suddenly remove his incapability. Nor does our acknowledgment of the limitations imposed by the mapping process suddenly disclose reality, as it might be for a nonhuman superknower, naked to our eyes. On the other hand, the model of ontology as a global mapping of totality tells us that, when all limits are recognized, there is always something to put on the map, and we know that the impossibility of putting everything that is known about the earth's surface on the same map does not prevent cartographers from making very useful and accurate maps of the world.

Just to keep from getting dizzy and vague in an area of thought that tends to overpower us by its sheer magnitude and multiplicity of possibilities, let us inspect the steps constituting the procedure of map making to see what they would involve in the case of totality mapping. Let us group them thus: (1) the scale; (2) the systematic distortion (grid and projection); (3) selective distortion, including its complement, the omissions; and (4) the content, or symbolic devices.

The Scale. The part of the mapping model most likely to cause despair at the very outset of any project of ontology is the scale, for the area of a map of totality must be *potentially* infinite and therefore not a measurable quantity at all, either temporally or spatially. This means that we must abandon any idea of ontology as a fixed and final description of the eternal and unchanging "structures of being" adopting an area that is expandable to accommodate the increase of knowledge and contractable to eliminate the accretions of mistakes and vain imaginations. There is nothing about the geographical mapping model to discourage us in the face of real changes taking place on the surface of the earth—continents rising and submerging, even the size and shape of the planet changing—and nothing to prevent us correcting mistaken notions about little-known lands. Man himself changes the surface of the earth and maps the changes he has made. Above all we must not a priori evaluate change as evil, and, like Plato, panic at the thought of a Heraclitean universe. Consider the sun—it has a changing surface, yet we can map the appearance and disappearance of sunspots, the eruption of prominences, electrical storms, and so on. The threat to knowledge posed by *rampant* Heracliteanism is less the problem of change than that of the *rate of change* relative to our nervous system's ability to notice or infer or record or explain the changes. We already know a great deal about changes that are too slow or too fast for anyone to perceive, thanks to our scientific "models," and we will learn more in the future, even though our brains should gradually change, through evolution. What Plato could not realize was that our nervous system was made to operate in the midst of changes and would atrophy in immobility—but then, he would probably consider the brain as it is understood today to be a bad imitation of something better.

The dotted line on the crude map of the matrix world (Figure 1*a*, p. 151) represents the expandability of the area of the individual consciousness, as well as the variable boundary line between whatever knowledge an individual may possess, and the impenetrable mystery that surrounds the maximum knowledge that the maximally informed hypothetical individual or collection of knowledge-pooling individuals might among them embrace. This dotted line is the knowledge analog of the total surface of the earth, to be mapped in any ontology; therefore inclusiveness is bound to be one of the standards by which we judge the relative merits of one ontology as against another. *Maximum inclusiveness*, although unachievable as a final goal, gives an irreversible direction to the ontological quest, imparting to the whole enterprise a cumulative character, not reducible to "more of the same." The modern lack of interest in this quest is probably due to the well-grounded suspicion that maximum inclusiveness would mean that ontology would have to be done over and over again, at any given time trying to comprehend as much of the past-present-future as that point in history made available. Who would want to undertake such a hopeless and unfinishable task! But the task is not impossible, for as long as the goal of maximum inclusiveness is not allowed to make any claims to omniscience, by the acknowledgment of the surrounding mystery, there is always something to put on the totality map, along with the *caveats* against finality. At any point in history, ontology would properly claim to be only a kind of trial-balance sheet of the inventory of regional mappings representing the best of the different kinds of knowledge up to that point, along with their various probability indications for the future. Even for the radical positivist, then, equipped with only two kinds of knowledge, logic and a physics-modeled science, there would be possible a trial summation at any point in history of the maximum knowledge available at the time, all else being relegated to the chaotic interplay of noncognitive emotion, accident, "blind" instinct, and so on.

Systematic Distortion (The Grid and The Projections). After the sliding scale, constantly readjusted to maximum inclusiveness of knowledge and maximum elimination of errors and vain imaginations, the next limitation to consider with regard to totality mapping is the systematic distortion that would inevitably be introduced by the fact of its being expressed in a verbal language, and originally in a particular language. Of course we must allow for the possibility that there may be a better language for ontology than words, such as music or ritual action; but assuming that it would be expressed in words, we know that no matter which language we chose to express it in, a systematic distortion would be introduced into our description of totality by the vocabulary, grammar, and syntax of that language. Therefore, it would be better right from the start to think of our model not as a globe, but as a map of the world projected on a flat surface, such as Mercator's map, or the many queer-shaped cutouts that can serve to show the total surface of the earth in a warped way. In this way we would both acknowledge and dispose of the objections from the antimetaphysical camp that ontology is a misuse of the verb "to be." Whose misuse is it? In what language? Of the estimated 5000 languages mentioned in the *Encyclopaedia Britannica*, relatively few have the verb "to be" in their grammars at all, and of these a much smaller number use the verb in the sense that the Greeks used the abstract noun *ousia*, either as the intelligible nature of a thing—a being—or as the intelligible nature of being-itself—the totality of *what is in reality*. But does this lack of a particular usage of "to be" prevent these other languages from talking about totality with other verbal devices? It has not so far. Here we must be guided by what was said at the end of Chapter Five regarding the check on the idiosyncratic distortion tolerable in any language provided by the need for communication, plus the "natural contours" of the perceivable world. In other words, we must not mistake the *code* for the *message*, and if ontology has any message concerning the totality, the message must be translatable into languages other than those that use "to be" in the Greek philosophical sense. We must help ourselves out by invoking the *referent*, and if totality is the referent when "being" is used in the ontological sense, then by the same token, any word or phrase used intentionally to refer to totality should be regarded as at least an attempt at an ontological way of speaking. Examples would be God, the All, the One, the Brahma, the immeasurable ocean of being, the Encompassing, the most inclusive reality, the Great White Father, the World Spirit, and "everything that is the case."

The problem, however, is not as simple as that. It is not only a question of finding verbal equivalents between any two languages (which even a well-programmed translating machine might do), but of taking account, in the translation, of the "frozen philosophy" that is hidden in the structure of every language, or at least family of languages. This is a problem of ancient lineage for Christian theology, because Christ spoke Aramaic, a variant of Hebrew, yet all his words and acts that have come to us were preserved in Greek, even though the frame of reference in which those words and acts occurred was that of a Hebrew community immersed in Old Testament ways of thinking. We have only to imagine how different the history of Christian doctrine might have been if Paul had gone east instead of west. In that case the first civilized pagans he would have come across would have been the Persians; Christianity would have had to define itself over against the religion of Ahuramazda and Ahriman; and Christian doctrine would have been cast, by the Early Persian Church Fathers, into the open and hidden philosophy encapsulated in the Persian language. But, in any case, a missionary religion would sooner or later have had to spread out from the Semitic homeland; and we should now think of its translation into Greek as the price that had to be paid for its conversion of the Hellenized Mediterranean world at such an early stage in its history. In the case of ontology, and going in the opposite direction, we cannot help wondering what Aristotle's *Metaphysics* would sound like in ancient Hebrew, or in Chinese, or in Nootka!

Fortunately for us, the professional linguists are taking some of the winds of doctrinal dispute out of the theological sails by exhibiting the frozen philosophy hidden in noun forms, verb forms, tenses, and spatio-temporal expressions of different languages. A pertinent example for our purpose is Thorlief Boman's *Hebrew Thought Compared with Greek*; a few examples from this work illustrate the difficulties that face the project of ontology because of the factor of systematic distortion. Boman notes how the verbs in Hebrew express the dynamic, activist, restless character of Israelite thinking, since even the "stative" verbs, those which indicate position, sitting, lying, or being in some condition, also designate a movement.

> Our analysis of the Hebrew verbs that express standing, sitting, lying, etc., teaches us that motionless and fixed being is for the Hebrews a non-entity; it does not exist for them. Only "being" which stands in inner relation with something active and moving is a reality to them. This could also be expressed: only movement (motion) has reality. To the extent that it concerned Hebrew thinking at all, static being is a being that has passed over into repose. (10)

Static being in its logical sense as *what* something is, is expressed by means of the noun clause, where both the subject and the predicate are substantives, thus eliminating the copula and various verbs of inaction: "If Jahveh (is) (true) God, follow him, but if Baal (is), then follow him." Being, expressed in the verb *hayah*, has the triple connotation of existence, becoming, and effecting, and it shifts its emphasis according to the context. Thus *hayah* of God is to act as God, to deal as God, and to carry into effect as God.

> From *hayah* we can understand what "being" was consciously or unconsciously for the Israelite; "being" is not something objective as it is for us and particularly for the Greeks, a datum at rest in itself. It is, however, quite erroneous to conclude from this that "being" is something subjective, evanescent and dependent upon us. The Israelites like all other ancient peoples were "outer-directed" and did not dissect their psychic life as modern man does. In that

sense, even to the Hebrew, "being" was something objective which existed independently of him and stood fast. The "being" of things and of the world as the totality of things was to him something living, active, and effective, a notion which, however, has nothing to do with primitive pan-psychism. It is correct to say in the case of Hebrew that "being" (e.g., the being contained in stative verbs) represents an inner activity which is best to be grasped by means of psychological analogies with human psychic life; with that we come to the heart of the matter. In the full Old Testament sense "being" is pre-eminently *personal being (Person-Sein)*. What does it mean that a person *is*? If we try to define that by means of the concepts of impersonal and objective thought, we have to grasp for "becoming" as well as for "being" and still fall far short of the objective. The person, on the other hand, is in movement and activity, which encompasses "being" as well as "becoming" and "acting," i.e., he *lives*; an inner, outgoing, objectively demonstrable activity of the organs and of consciousness is characteristic of the person. Personal being is a being *sui generis* and therefore cannot be expressed in terms which are grounded in impersonal and objective thinking. A system of thought in categories that stem from personal being, *mutatis mutandis*, does not do justice to objective and inanimate reality. (11)

Professor Boman's attitude is conciliatory. I can summarize his argument best by making two lists of words that describe the idiosyncratic tendencies of the two ways of thinking hidden in the two languages, and then giving his conclusions.

The Greek Way (also Indo-European)	*The Hebrew Way (also Semitic)*
Seeing, clear image, idea	Hearing, feeling, response
Space-oriented	Time-oriented
Harmonious, orderly	Powerful, effective
Restful, cool	Warm, emotional
Architectonic	Unsystematic
Geometrical, sculptural, formal	Impressionistic, sensory, bodily
Unhistorical	Historical
Impersonal, objective	Personal, dramatic
Abstract, universal	Concrete, collective

From these differences in the two languages, Professor Boman concludes that the Greeks were providentially destined to reach their highest achievements in the fields of logical thinking, philosophy, and science, whereas the Hebrews were destined to achieve their utmost in deep psychological understanding, morality, and religion.

But here we must ask whether Boman is comparing comparable things. Should he not have compared the Bible's psychological understanding, morality, and religion with the works of Homer and Hesiod, with the tragedies and comedies, the festivals, and the accounts of the Persian and Peloponnesian wars? Rather than with Plato and Aristotle? According to my own analysis of Ordinary Language as the verbal mapping of the everyday world, all languages, just to have survival value for their users, must contain some elements from both of Boman's columns. Which of these are regarded as more important is a matter of cultural–historical selection. Is it not possible that the philosophical Greeks may have been encouraged by the very chaotic conduct of the gods and goddesses in their mythology to seek some more orderly explanation of the phenomena in the visual world, whereas for the Hebrews, conversely, the unifying effect of the One God discouraged scientific curiosity about the structure of His "handiwork"? And as against Benjamin Lee Whorf's "thought-molding-by-the-native-tongue" theory, did not many Greek-speakers become poets and dramatists instead of scientists, and did not many Hebrew-speakers, later on, become scientists, logicians, and mathematicians? Who is to say how much of this difference between the Hebrews and the Greeks was temperamental, how much intellectual; and if temperamental, how much hereditary, how much cultural; how much *deposited in*, how much *transmitted by*, their language and literature in response to their history? But I am only using Boman as an example of the difficulties that must be faced by ontologists of the future.

The shrinking of the initial geographical isolation that brought about such fantastic proliferation of languages leads us to expect a gradual reduction of the number of distinct "native" languages. At the same time, for those languages which do survive into the "one world" period, there will probably be a new proliferation to take care of the avalanche of knowledge as it spreads to all peoples, this time by the addition of "specialized" languages to "ordinary" languages. We can foresee a great future for the translating business, with all the arts and wiles and strategies and failures that it will involve. Ontologists (or metaphysicians) of the future, not being so wedded to any one code that they will ignore a striking aptness of expression for a particular message in some other code, may help themselves out by judicious "adoption." Thereby, gradually, future ontologists may evolve a metaphysical *lingua franca*; indeed, the strategy is already evident in the use in English prose of untranslated phrases from Heidegger's German ontology. Furthermore (since we are speculating), if our type of civilization survives, the Tower of Babel consisting of "native" languages will gradually be replaced by a Tower of Babel consisting of the "specialized" languages, and a new breed of *translators* will be required: those who practice the art of decoding and recoding in order to pass messages from one specialized-language subculture to another. (Overheard, economist to theologian: "Is it really true that you can't get along without a word like *metanoia?*") Ordinary Language will still be taught to children and used in everyday life, but it will be tremendously expanded by "technical slang" on the one hand and by worldwide historical and literary references on the other. How grammar and syntax will be affected by such a development is anybody's guess, but they will surely be better understood than now as the structural elements of a code.

Selective Distortion (Including the Omissions). The foregoing bit of projective imagining about the linguistic future already encroaches upon the most difficult of all ontological problems—the question of what to include and what to exclude in any attempt at totality mapping. "It is impossible to put everything on the same map and still have a map instead of chaos," holds true with a vengeance for ontology, precisely because of the scale factor already discussed; that is, the demand for maximum inclusiveness. But how we combine maximum inclusiveness with the selective distortions and omissions that are necessary if the map is to be *intelligible*, let alone useful? The geographical cartographer, before he begins, asks who is going to use the map for what ends. He also determines what will be important and what will be safely negligible for those purposes. Well then, what is the purpose of ontology? What can you use it for? And what can you learn from it that you could not learn far better from one of the regional maps, which do not make such grandiose claims and are likely to be far more informative? The purpose cannot be just to make an inventory of all available knowledge up to the present time. Many of our liberal arts colleges offer to the youthful student, prior to his specialization, a series of survey courses promising him just that: an inventory of what there is to choose from, under such rubrics as the physical sciences, the biological sciences, the social studies, and the humanities. But it has not been noticed that the students issue from this survey as convinced (or even budding) ontologists. They are more likely to finish the semester confused, overawed, and prudently determined to waste no more time on anything but that for which they are fitted by heredity and situation in life. Still, it is an invaluable exercise for expanding the consciousness, and one that should be a minimal prerequisite for any would-be ontologist. The purpose of ontology is to make sense, human sense, to make some sort of articulated, interrelated pattern, picture, gestalt, or map, out of the whole; and one cannot hope to do that if he does not have at least a minimal knowledge of what *could* claim inclusion in that whole. Furthermore, the survey has a historical, developmental side, both from the scientific and from the humanistic aspect; and many of the items in it will strike the modern ontologist either as errors or as vain imaginations that need not be perpetuated, although it should perhaps be explained why they happened, if only to maintain continuity with the past and to make sense out the denotations and connotations of our present words. And how else is he to evaluate what he wishes to keep as knowledge and what he wishes to get rid of or explain as errors and vain imaginations, except by means of some theory of knowledge that he brings with him to the job?

There is no point in arguing over priorities as between epistemology and ontology, since any theory of man, insofar as it tries to be the most inclusive view of man, will betray an ontological outlook, and any ontology will have to use some sort of epistemology to decide what to include or exclude from its totality pattern. The only thing to do is to lay one's cards on the table, as we did when we said (in Appendix Two) that the congruent view of man used in this epistemology would be a biblically derived doctrine of the wholeness of man.

But here the basic "model" or "construct" comes in—to help us to decide what is congruent with what. If we ask what Plato's basic model of knowledge was, we find an ambiguous shifting between two models (if at this distance of language and time we can claim to understand what he was trying to say.) As long as he was talking about the mind in relation to the eternal Ideas and Forms, we have the following basic model: knowledge is a kind of *remembering* (from the prenatal journey around the heavens). Whereas when Plato refers to the present world of change and becoming, the basic model is that knowing is a kind of eating—the mind is thought of as a kind of intellectual stomach that takes in, becomes united with, and *becomes like* the intellectual food it eats. (Note that Eros is in want, hungry.) How else is it that, in the *Republic*, it is assumed that the man with the better definition of Justice in his head will *become* more just than the man with the worse definition or than him with none at all? Now, aside from the fact that even the real stomach does not become like the food receives, the eating model for knowledge leads to the absurdity that the man with the best definition of Justice in his head, by the simple logic of opposition, also has the best definition of Injustice in his head—so why does he not *become* just as much unjust as just? And similarly for other definitions in terms of polar opposites. On the other hand, in our model of the mind as a receiver and evaluator of, and responder to, messages, the mind needs both the positive and negative definitions (brief descriptions) to tell it what to strive for and what to avoid (stop-and-go signals). So the basic model must not lead to absurdity, or else it must be "doctored up" so that it does not. But the basic model must also try to bring about, not abstract unity nor universality, but a unifying congruence between the various kinds of activity that this kind of animal actually must undertake in the variety of life situations it must deal with in order to survive and to make that survival worthwhile, for the same mind, not a variety of minds, is used in each activity.

The process of comparing one's basic epistemological model with past and present ontologies is obviously going to lead into the study of religions as well as philosophies, since at least the "higher," more universal, religions are most inclusive views of reality that have already been evaluated according to the criterion of what is most important for man. Such "comparative" studies as the history of religions, the phenomenology, the sociology, and the psychology of religions, are possible only because we are able to "translate" innumerable concrete mythologies into a few recognizably *comparable* philosophical or psychological terms, such as the God, the gods, this life as against the afterlife, change as against no change—in other words, basically ontological terms. "Comparative religion" is really comparative ontology, not excluding polytheistic, nontheistic, and antitheistic ontologies. For this reason it can be said that everyone secretly or unwittingly worships something as if it were God (i.e., as his God-surrogate). The conditions of human existence force everyone to have his most inclusive view of reality and within that his evaluation of what is most important for man, or for himself. For this reason we sometimes call him *homo religiosus*, but not, be it noted, in a congratulatory sense! Of course he would be better off if he could afford to remain in a state of suspended judgment until the end of time, when it will perhaps be more evident what is the most inclusive view and what is most important for man—but he is not given that privilege, if privilege it would be, to view things *sub specie aeternitatis*.

Within this wider context, then, I can give an example (but I cannot go into comparative religion) of some ontological viewpoints with which the epistemological model of this essay would be more, or less, congruent. It would be least congruent with any philosophy or religion that emphasizes an attitude of passiveness or quietism or fatalism with regard to the world as it is. In Chapter Two we had to draw the knowledge line between the plants and the animals because the plants were too passive and did not have any means of either receiving or responding to messages by means of actions toward, or away from, the beneficial or harmful situation; moreover, they had no way of modifying future behavior according to the feedback signals of past experience. This does not mean that passive contemplation, especially of the "All," might not be a much-needed temporary rest from the strenuousness and "smallness" of living, but only that if it were regarded as the most important attitude for man, it would make it impossible to regard any content of consciousness as knowledge, and any action as mistaken or correctable or right or wrong. The original Hindu religion is amazingly consistent in this, for the "drop of water" tries, not to change the "Ocean of Being," but to disappear in it. Accordingly, life in the world of change is not knowledge but illusion. (Certain similarities to Plato are striking, even to the four functional castes of society and the reincarnation of souls.) However, as soon as the first reformer, the Buddha, came on the scene, he spoiled this fine consistency by recommending a course of *action* (right behavior), complete with ethical prescriptions for getting to Nirvana, thereby using experience to modify the animal value system in order to ameliorate human existence; and with this Buddhist view of action, the present epistemological model would be somewhat more congruent. The deep emotional contradiction between complete detachment and the requirement of compassion then led historically to the subdivision into the Theravada, or ascetic and world-fleeing type and the Mahayana or compassionate type of Buddhism; there would be an even greater congruence of our model with the latter, insofar as it involves self-transcendence for the sake of others (see Chapter Six). Surely there is a spiritual analog between the Christian who lives "in the world but not of it," and the Bodhisattva who, "having attained the goal of purification and emancipation, refuses to enter Nirvana, out of devoted love for those who still remain behind and a consuming zeal to help them. He postpones his own entrance into perfect bliss because his sense of spiritual oneness with others leads him to prefer to wait with them and lovingly serve them until all are ready to enter together." (12)

This example only begins to illustrate that the present model does not necessarily indicate that there will some day be only one universal religion, or only one true ontology, or even that such an agreement would be desirable. There are many stages on life's way, many decisions to be made, commitments to be enacted, and risks to be taken in situations as yet unforeseeable, and many experiences vouchsafed to some and denied to others; we should not expect any individual simply to adopt a philosophy or religion by majority rule, when its ontological outlook is unrecognizable to him or is contrary to his own experience of life up to that moment. But we should expect that the several different types that appear to be "live options" of ontological orientation at any one time would be more willing than not to learn from one another, especially if they understand the tension between maximum inclusiveness and maximum importance brought about by the necessity for selective distortion and omissions. Thus a good grounding in the writings of the classical mystics of the higher religions might not be a bad prerequisite for ontologists of the future—those in this "grass-roots" tradition of religious experience at least try to bring one to a *pause* before the unfathomable mystery of the totality, even though they proceed to interpret it in different ways and to recommend different attitudes and actions toward it. The Protestant tradition has been altogether too suspicious of mysticism, precisely because of the NeoPlatonic form it has taken in the Catholic tradition (where it is thought to be too erotic, too selfish, too world-abandoning, and too much in danger of self-deification and of by-passing Christ altogether). Indeed St. Augustine himself complained that he would never have learned charity from the NeoPlatonists. But could he not have learned something of charity from the Bodhisattvas of Mahayana Buddhism? (13) If we may speak of a Biblical ontology—and perhaps some day we may (or at least of an ontological outlook betrayed in the writings of psalmist and prophet)—it is not hard to see that the unfathomable mystery is the transcendent aspect of the Creator God of Genesis: His wholly-otherness, His hiddenness and remoteness from man, the Uncategorizable, as He or It might be to Himself or Itself (but even He and It do not apply). To this aspect of God, one might well afford to be agnostic. But in and through the creation, in and through human knowledge, he shows himself to man, and he is known as ultimately the Sender of the messages and the Receiver of the responses. Or in Buber's dialogic philosophy, He is the Thou who addresses me, and to whom I, as long as I live, in one way or another, in fact do respond.

The Content, or Symbolic Devices. As a writer, the ontologist will be in the same position as all writers; namely, he will have to choose his words, invent his verbal devices, or pursue his verbal strategies, so that he can serve the expressive and communicative purpose of the kind of literature he will be creating. Everything that was said in Chapter Three about the various knowledge-enhancing and knowledge-obstructing functions of the imagination, insofar as it is true, now applies to him as a writer. In relation to his specific project, totality mapping, he will have to use or exercise imaging-as-if-perceiving with regard to past, present, and future; also impressionistic imaging, projective imagining, empathic imagining, and expressive imagining; and he must try to avoid vain imaginations. His entire "native" language will be at his disposal and perhaps several others, and not only the verbal devices of ordinary language but also the symbolic devices invented for use in the specialized branches of knowledge we have called regional mappings. This embarrassment of riches will constantly remind him that his job is to select or to invent, but always in relation to the appropriateness of the words for those aspects of the subject matter he has evaluated as most important within the most inclusive view. He will have to keep in mind an imagined reader, who will be first of all himself exercising the critical function. But he must also consider a more generalized reader who will be imagined to have a certain equipment in the technical language of, say, philosophy, psychology, theology, and ancient and modern history. He will have to make the important decision of whether to invent neologisms or whether to try to assign new or specialized contexts and usages for existing words. He will have to be acquainted with "the present situation" in philosophy, so that he can avoid repetition and say something that needs to be said—exactly like the writer about knowledge in any other field. As instructive examples of the variety of vocabularies adopted or adapted by some contemporary writers in the range of near-totality mappings, compare Whitehead, Tillich, Santayana, Cassirer, Heidegger, Marcel, Toynbee, Teilhard de Chardin . . .

Let us for brevity's sake refer to the available list of symbolic devices as the range from mathematics to mythology, covering in this range all degrees of abstractness and concreteness, objectivity and subjectivity, certitude and incertitude, definiteness and vagueness, provability and unprovability. From the standpoint of the present epistemology, *all these symbolic devices are equally anthropomorphic*, precisely because they are products of the way the human brain works in its niche in the impenetrable mystery, as against the way a fish brain, or a butterfly brain works (such "symbolic devices" would properly be called ichthyomorphic or lepidopteromorphic). There must, therefore, be a reason why some philosophers, scientists, poets ("modern" men) want to use the term "anthropomorphic" in either a pejorative or a commendatory sense. This reason becomes more evident in the special project of ontology than in regional mappings, which do not claim to "cover" totality. In the case of abstractness versus concreteness, we know that the built-in generality of abstractness is gained at the expense of concreteness, and that concreteness is gained at the expense of generality; but this comes out more clearly when generality is extended to universality (i.e., made applicable to totality). The "concrete universal" of Hegel is an impossibility for the human type of knowledge, and once we understand that knowledge is for the benefit of man, we can dismiss Hegel's claim as a vain imagination or as one of the endless pairs of opposites that can be spun out endlessly about the impenetrable mystery. (14) Our "survey courses," listed earlier, progress from the abstract to the concrete, the physical sciences leaving out the biological, social, and human knowledge for the sake of *applying to them all*, the human knowledge gaining concreteness and definiteness *at the expense* of applying to them all. Most people survive by having a sort of common-sense, unarticulated (intuitive) understanding of this situation, but when ill-digested philosophy becomes popularized, we witness the war between the engineers and the poets, the engineers calling the poets liars because their "truths" cannot be proved, the poets calling the engineers liars because *their* "truths" have left out the most important part. (Remember Yeats's "holy hatred of science.") Even in a friendlier mood, we get only scientific humanism, the scientist saying to the humanist "Tell us what your 'values' are, and we will show you how to reach them," and the humanist complaining "But I do not recognize myself on your drawing board!" The ontologist's job is to do justice to both.

The engineers did not achieve their universality along with their exactness, certitude, and predictability, by saying things like "all is one," "all is many," and "everything is earth, water, air and fire." Rather, they made a model of a very small area of experience in terms of a few abstractable variables, and generalized it to see how far it would go before it collided with the facts and had to be remodeled. Then they eliminated all the facts that made no difference to the model—and so the process went. The poets did not get their concreteness and definiteness out of the realm of privacy and opaqueness (commotion) by uttering cries as children do or by throwing tantrums, but by using the resources of language (which already contain a degree of abstraction in their communicable images, metaphors, similes, action-and-attitude words, and their hosts of connotations) for mapping the interior landscapes of consciousness. The ontologist can select from all these elements because, as we said in Chapter Five, "it is precisely the advantage of maps that they can use many combinations of pictorial and abstract devices to serve a variety of purposes."

But there is no getting around the failure of ontologists (metaphysicians) of the past to understand the delusive character of abstraction, however much we may exonerate them by the condition of knowledge in their time. Therefore, ontologists of the future, whether they are starting from scratch or refurbishing an Old Master, must try to learn what they can from the Positivist attack on metaphysics as reflected in its ridicule: News from Nowhere; Exalted Nonsense; Bigthink. The news must be from everywhere; there must be different kinds of sense-making, including the most depressing; there must be no unjustified claims for human knowledge resulting from generalizing or extending a concept far beyond its original area of development and producing vacuousness, false analogies, and spurious universality. Thus when Whitehead extends the concept of "experience" clear through to the inanimate part of the world, physicists should object, "animistic physics!" Similarly, when Teilhard de Chardin carries "the within of things" well below the inanimate level. Or when Tillich, using Heidegger, says that Being has the character of self-world correlation. *Some beings* have this character, and some do not. But if Tillich was using Being as a totality symbol, a stand-for-God word, he should have included the transcendence that leads to Impenetrable Mystery. Of course Whitehead could be corrected and updated, as also could Heidegger. In Part Two, Chapter Eleven, I suggest that Heidegger's Being-in-the-world should be corrected to Becoming-toward-totality-through-imaginative-reciprocal-interaction-with-an-expanding-world. Perhaps there should be a conference between Whiteheadians and Heideggerians, to see if they can learn something from each other. The latter might not feel so "thrown" into the world-nothingness if they paid more attention to the long process by which they got here.

The totality-mapping model for ontology (metaphysics) makes unjustifiable claims for human knowledge simply unnecessary. Large areas can quite legitimately be left as *terrae incognitae*, whereas others sketchily known can be filled in or corrected. Regarding such questions as the sentience of plants or the extraterrestrial origin of life from organic molecules in interstellar space—is it not more honest just to leave them open than to people the universe with nonexistent "somethings" for the sake of premature order, unity, continuity, and harmony? Let those who wish to talk to flowers consider that if plants had feelings, every grass-cutting would be a massacre. I stand ready to be corrected on the question of protosentience in plants, but I had to *delimit* my subject matter, protoknowledge, precisely to guard against false analogies. Even if it turns out that plants have some sort of passive protosentience, their immobility prevents them from approaching the beneficial and avoiding the harmful situation, thus precluding protosentience as action guidance, or protoknowledge.

When it comes to action guidance on the human level of consciousness, the delusiveness of abstractions is evident because both sides in a political power struggle, or even in a war, are quite capable of using the same words as slogans: *for* Freedom! Democracy! Justice! Equality! *Against* Slavery! Oppression! Imperialism! War! "Make love, not war," a favorite slogan of the young, prettily glosses over one thing: making the kind of love they have in mind is one of the chief causes of war, by way of overpopulation, overcompetitiveness, and overacquisitiveness on behalf of the loved ones. Would they be willing to follow St. Francis's prayer, which spells out the kind of love that would be an "instrument of peace"?

Avoid Glittering Generalities! ought to be the chief action-guidance slogan for ontologists of the future, taken by them as a monastic vow. It is not easy to obey, however, given the demand for maximum inclusiveness. Spell out what is meant: Freedom from what? For what? For whom? At whose expense? Those who are aware that glittering generalities are both bad science and bad poetry (see Chapter Eight) will maintain a healthy ambivalence toward language, especially the dangers of abstraction in the human field of meaning. Hugh Kenner wrote it for poets, but ontologists should also beware: "Language is a Trojan Horse by which the universe gets into the mind. The business of the artist is to be constantly aware that the horse houses armed warriors, even while admitting it to his mental citadel. He has then a chance of winning them over, ..."(15) Every time we write the word "man" we should blush, knowing that each member of the category is a unique version of the inherited animal value system (the collective unconscious?), that each has a unique history superimposed, and even identical twins do not go through the same motions from the instant of birth.

Be on the lookout for false analogies! Jumping from the atom to man and back again with concepts like "free will," "determinism," and "natural law" is a great proliferator of false analogies, ignoring entirely the intervening biological, cultural, and historical episodes. Since it is not easy, even in the case of man, to decide what these ideas mean in what specific contexts, why foist them on the whole universe? The model maker is suffering from modelitis as is also the careless user of the word "creative" to describe evolution. Suggested analogies between biological form evolution and conscious human form creation must be treated with the utmost circumspection. The latter involves directed curiosity, persistent attention, all the forms of disciplined imagination we studied in Chapter Three and the avoidance of vain imaginations. The *theory* of evolution is itself one of the most astounding examples of conscious human form creation, and it certainly made order out of the chaos of animate forms for the human type of mind. But it also required, on the part of its contributors, a high degree of freedom *from,* and management *of,* the animal value system in its service of immediate survival needs. Does *it* have any survival value for man? Not if we become its victims by false analogies.

Finally, *shun reductive formulas!* Sentences beginning with "Man is nothing but—" will simply close the ontologist's mind to news from everywhere, and if he needs the kind of *certitude* that comes from uttering them, he had better get into a different line of business. Treating regional mappings *as if they were* totality mappings is cultural provincialism of the worst kind. On the academic level it leads to intellectual tribalism, the inability or refusal to communicate outside one's own discipline, and the cozy solidarity of the specialized-language group. But this is where we came in.

Have I made things too difficult? Surely anyone who took these three guidelines in strict seriousness would be reduced to utter silence, like a mystic of the *via negativa*. Or like Wittgenstein when he said, "Whereof one cannot speak, thereof one must be silent." Or the prophet who exclaimed, "God is in his holy temple. Let all the earth keep silence before him!"

REFERENCES

1. Aristotle, *Metaphysics*, Ed. and transl. by John Warrington. Introduction by Sir David Ross. Everyman's Library Edition, p. 55. Published by E. P. Dutton & Co. Inc., New York, 1956, and used with their permission. Also with permission of J. M. Dent & Sons Ltd., London.
2. G. M. A. Grube, *Plato's Thought*, Beacon Press, Boston, 1958, p. 1.
3. I. M. Crombie, *Plato's Doctrines, Vol. II, Plato on Knowledge and Reality,* Humanities Press, Inc., New York, 1963, p. 11. Also with permission of Routledge & Kegan Paul, London.
4. I. M. Crombie, *Plato's Doctrines*, p. 26.
5. Plato, "The Symposium," From *Great Dialogues of Plato,* as Translated by W. H. D. Rouse. Copyright © by John Clive Graves Rouse. Reprinted by Arrangement with The New American Library of World Literature, Inc., New York, 1956, pp. 104–106.
6. Aristotle, *Metaphysics*, p. 287.
7. Aristotle, *Metaphysics*, p. 57.
8. Aristotle, *Metaphysics*, pp. 395–396.
9. Aristotle, *Metaphysics*, p. 346.
10. Thorlief Boman, *Hebrew Thought Compared with Greek*, Westminster Press, Philadelphia, 1960, p. 31. Reprinted also with permission of SCM Press, Ltd., London.
11. Boman, *Hebrew Thought*, pp. 45–46.
12. E. A. Burtt, *The Teachings of the Compassionate Buddha*, Edited with a Commentary, A New American Library Mentor Book, New

York, 1955, p. 127. In particular, read pp. 161ff, for the retroactive effect of the Bodhisattva religious ideal on the concept of Nirvana.

13. F. C. Happold, *Mysticism, A Study and an Anthology*, Penguin Books, Baltimore, Md., 1963. This book presents excellent brief introductions relating mysticism to the present scientific view of the world.
14. See Nicolas of Cusa's *coincidentio oppositorum* in F. C. Happold, *op. cit.*, p. 302.
15. Hugh Kenner, in G. Seon, Ed., *James Joyce, Two Decades of Criticism*, Vanguard Press, New York, 1948, p. 144.

PART TWO

The Applications

Eight Philosophical Analysis and Linguistic Philosophy

The movement of philosophical analysis and linguistic philosophy arose in response to the linguistic impasse left by Logical Positivism, namely: if the only propositions that made sense were the propositions asserted by scientists, what on earth was the rest of the world talking about? The Positivists themselves had consigned all nonscientific uses of language to the category "emotive," and by their dichotomy of cognitive/emotive, they emphasized their conviction that the category "emotive" must mean noncognitive. After that, any epistemology that wished to be more pluralistic would have to challenge this dichotomy and show not only how emotive could also be cognitive, but how and why various mixtures of factual and evaluative knowledge were needed to interpret the whole of human experience. One would have to construct a spectrum of the mixtures to act as a bridge between the extremes.

Table 1 shows the kinds of mapping strategies and mapping devices used in the different kinds of knowledge. I have arranged them in a spectrum that stretches between the extremes that seem to be most opposite to each other—good science and good poetry. The term "good" should not be interpreted to mean that there exists, in some heaven of ideal mod-

TABLE 1 THE RANGE OF MAPPING STRATEGIES AND MAPPING DEVICES USED IN THE DIFFERENT KINDS OF KNOWLEDGE

Good Science (makes bad poetry)	Mixtures of Mediocre Science and Mediocre Poetry (good science and good poetry do not mix)	Good Poetry (makes bad science)
Overall strategy is to eliminate the concrete in order to reduce the number of variables, increase control, and enable prediction in terms of abstracted variables and constants in a model. The ideal is mathematical description of relations in a model and verification of predictions from the model under controlled conditions. Limitations are too many variables, not enough controls, imponderables and unrepeatables, pulling back to middle section ⟶	Ordinary Language used in everyday experience History of "old" philosophy Logic Theory of valuation Ethics Aesthetics Metaphysics (ontology) Theology Mythology and sacred writings The "wordy" sciences Economics Sociology Anthropology	Overall strategy is to avoid abstractions in order to focus on what good science has eliminated: full, concrete, individual richness of experience. Verbal abstractions used not so much as concepts, but like images, for their emotive overtones, metaphorical reverberations, multiple meanings, etc., in order to condense much into few words. Limitations are the fantastic, the unrecognizable, the confused, the chaotic, the vacuous, pulling back to middle section

Aim is hypotheses that are exact, quantifiable, fitting in with others already verified, leading to new, repeatable experiments.	Psychology Phenomenology The reconstructive studies Cosmology Paleontology, including evolution Archeology History Projective studies Utopias City planning World organization Near-future problematics Pollution, population, exhaustion of resources, nuclear war, space exploration, etc.	← By extension, the nonverbal arts, but with suitable adjustment to their sensory modalities. Aim is clarified, unique configurations, embodying or reenacting feeling, leading beyond themselves.

els, a standard of ideal science and ideal poetry—no, because the *ideal* would be that no distinction between science and poetry would have to be made. "Good" simply means the best we can do with the means at our disposal, and the arrangement emphasizes again that good science must be compared with bad science, good poetry with bad poetry (not science with poetry). Science and poetry are kinds of knowledge, but they have different knowledge jobs to do. It is the nearly opposite kinds of mapping devices that each is forced by its subject-matter to make use of—what makes good science makes bad poetry, and what makes good poetry makes bad science—that indicates precisely that the two are most nearly "opposite" to each other.

Between the two extremes are the kinds of knowledge that, for various reasons, constitute mixtures of the two extremes in a far-from-good state. Good science and good poetry, because of their opposing strategies and devices, just do not mix very well, and so I have decided to call the middle section mixtures of mediocre science and mediocre poetry, shading off into mixtures of bad, inept, or misguided, science and poetry. We all begin as children, in the middle section, starting off with ordinary language. The human race itself started off with Ordinary Language under the pressure of primitive survival needs; the two extremes can be thought of as specializations away from the middle, historically developed under the more "civilized" needs for better control over nature, for social order and solidarity, and for individual expression and communication. In case any crypto-Hegelians are reading this, the middle section should *not* be thought of as the synthesis of the extremes, nor should the extremes be thought of as opposites in the positive–negative, thesis–antithesis, sense. Good science and good poetry are *equally positive*, but about different portions of experience. It would have been just as possible to call the "opposite" of science music or the dance or ceremonial, but in Table 1 I am sticking to verbal devices.

Table 1 can be regarded as the epistemological "bridge" between the early positivism and the later linguistic analysis. It is also a bridge between the sciences and the arts, however, and it is not just one bridge but a multiple bridge—many strands or connections can be made across the middle section where the far-from-good mixtures are found. These strands or connections will depend on the degree to which we recognize that projective and expressive imagining and imaging-as-if-perceiving and evaluating are involved in making scientific models and devising predictions and verifications based on these models, and, on the other side, the degree to which we recognize that conceptual thinking is involved in the structure and functioning of poetic "models," the finished literary works. We could say, in a way, that good science and good poetry are complementary: each one deals best with what the other has left out. But each one encounters limitations in the subject matter itself (which is *all* of experience) that tend to push it back toward the middle section. As science tries to deal with more complex subject matter, its limitations are revealed in the form of too many variables and not enough controls (as in medicine or sociology). As poetry tries to cover larger areas of experience in concrete terms, its limitations appear in the form of the fantastic, the unrecognizable, the confused, and the vacuous. These two sets of limitations may be thought of as two different ways in which chaos makes its reappearance when we push our ways of knowing to their limits, but they are also reminders of the ultimate impenetrable mystery always in the background of all our attempts to make order out of chaos. The Mystery (surely we may capitalize this omnipresent ultimate in our lives) is not itself either order or chaos, but appears to us as a chaos that is pregnant with order—an order that human knowledge can bring to birth one fragment, or layer, at a time.

Our essay has shown why the early logical positivists were certainly justified in trying to get a "better language" for physics. "Better" meant divesting the physicists' words—space, time, matter, energy,—of their Ordinary Language denotations and connotations (sometimes animistic, psychological, or humanistic) and also of their metaphysical-ontological in-the-totality connotations, and confining their connotations and denotations strictly to the model under investigation. The physicists themselves never got into any trouble when they did this—it was when the logicians came to their rescue that the wild surmises began. Logicians are driven by a passion for neatness, simplicity, and universality; and if it seemed to them at the turn of the century that the scientific statements were the ones most often making this neat and simple and universal kind of sense, they argued that all sense-making must be of the scientific kind and all other-than-scientific attempts at sense-making must be placed under suspicion of nonsense-making. However, it would be too easy to make the logicians the villains of our piece; it would be misguided too, for actually they are the unsung heroes of the epistemological quest. By deliberately making the effort to find the never-never land of the "ideal" language, they have demonstrated that no such country exists. The *Tractatus* may be regarded as the blueprint or map of this ideal country, while the *Philosophical Investigations* may now be seen as the log book of Wittgenstein's voyage of discovery into the actual linguistic world, a voyage on which much was learned besides the nonexistence of the ideal country. For example, the later Wittgenstein came to realize that the logician's supposed independence of "mere" empirical facts led to absurdity. The theory of logical atomism had demanded that, since complex propositions could be analyzed into atomic propositions, there *must* exist atomic facts or "simple objects" to correspond to them. Norman Malcom tells us in *Ludwig Wittgenstein: A Memoir*,

I asked Wittgenstein whether, when he wrote the *Tractatus*, he had ever decided upon any thing as an *example* of a "simple Object." His reply was that at that time his thought had been that he was a *logician*; and that it was not his business, as a logician, to try to decide whether this thing or that was a simple thing or a complex thing, that being a purely *empirical* matter! It was clear that he regarded his former opinion as absurd. (1)

Not any more absurd than if today we received a set of dot-and-dash signals from a distant planet and our logicians insisted, purely from analyzing the structure of this code, that there must exist, on that distant planet, some "simple objects" corresponding to the dot and the dash.

The proliferation of technical jargons and their partial adoption into Ordinary Language shows how the other sciences are also looking for a "better language" for their purposes, and it makes one wonder how it is possible, as some of the linguistic philosophers claim, that in Ordinary Language "everything is in order just as it is"! For if we move over to the other extreme of the spectrum of mapping devices, we find that poets are just as dissatisfied with Ordinary Language as are the scientists. They, however, do not have the option of using technical jargon. They find the Ordinary Language dull, bloated, platitudinous, pompous, fuzzy, full of overused images, peurile sentiments, trite metaphors, and emotionally inflated abstractions; but their only option is to use the *same* words in new combinations that will eliminate these betraying marks of bad poetry. After all, Ordinary Language was not designed for the purposes of good poetry any more than it was designed for the purposes of good science: it was designed for the purpose of producing those signals, attitudes, and actions necessary for the survival of the tribe.

In view of the great variety of functions that languages are expected to perform, and the potentially infinite language-eliciting situations, what are we now to think of the Positivists' Verifiability Criterion of meaning? First, we must credit Wittgenstein, even as early as the *Tractatus*, for having put his finger on the crucial spot in this criterion; namely, in order to have meaning, a statement must contain a reference to something observable in the world. He saw that even for science such a reference was not enough to give a statement meaning. Must there not be meaning in a proposition even before anyone can try to verify it, and must that meaning not remain there even after it is disverified? Nobody can verify a jumble of words, even if every word in the jumble contains a "reference to observables." The proposition "Unicorns exist" does not become *meaningless* because so far no one has discovered the beast described in legend as a horselike animal with a horn in the middle of its forehead. Wittgenstein would have said that "Unicorns exist" is the picture of a possible state of affairs, and remains that, even if no unicorns are observed to exist. Perhaps he did not realize that the realm of possibility is so much greater than "everything that is the case" and that it is imagination that is the organ for its exploration. To invent the category "unicorns" requires imagination of the projective kind, and to use it in symbol systems such as heraldry, painting, poetry, or mythology, requires imagination of the expressive and empathic kinds. And where would scientific theory be if for example Darwin had used verifiability as his sole criterion of meaning? And yet it was Wittgenstein himself who said, in one of his many parenthetical asides "(Uttering a word is like striking a note on the keyboard of the imagination)." (2) But he chose not to follow it up.

Of course we must continue to demand the highest degree of verifiability in statements that claim to be factual or to be based on "the facts." But verifiability by whom? At what point in time? Are we talking about the human mind in the present historical situation or about a race of superknowers in some superhistorical future found only in science fiction? What would count now, today, as verifiability in statements about specific historical events, even those happening "in front of our noses"? About desirable political arrangements? About preferable social policies? About the virtues of products advertised on television? About harmful or beneficial futures proposed for the human race? Somehow we feel that there *ought* to be some degree of verification possible in the thousands of claims and counterclaims we have to plow through to make sense out of our daily lives. But we tend to forget that in order to reduce the number of variables and increase the number of controls necessary for the verification of even the simplest relationships in nature, *the scientist's laboratory had to be invented.* History is a miserable laboratory, and supposedly simple everyday life situations do not remotely resemble the conditions in a laboratory, as the existentialists never tire of reminding us. Yet our very knowledge of what verification in the laboratory consists of places a kind of reproach against everyday life, a reproach that prescientific men surely did not feel. They too had to make important decisions, based on imponderables, unpredictables, implausibilities of inextricable complexity, and evidence that was unverifiable at the time, although it might be some day—when it would be too late to undo the effects of the decision. Man *must* act—and precisely in situations where his powers of reason tell him that he cannot act on the basis of reasoning powers stretched to the utmost—that is how Kierkegaard put it, and that is the now famous "existential situation" that has been labeled in the literature as "absurd." Prescientific men certainly thought life was difficult, but not that it was *absurd*, for they did not have the scientific criterion of the conditions of verifiability to compare it with.

The "games" model of languages is an unconscious or indirect underwriting of the view that human existence is absurd. The very word "games" carries connotations not only of the contrivance and arbitrariness of invented rules, but also of frivolous, pointless, time-killing, but pleasant, activity. Now of course there is a proper place in life for playing games (*Homo ludens*), but the word first appeared in *science* when the physicists who wished to become disengaged from metaphysics decided to drop the word "reality" (which, they said, had no more than honorific value) and to concentrate on making the "game" of science internally consistent by tightening the rules. Wittgenstein, who saw correctly that languages operate by following rules, would surely not have claimed that all following of rules constitutes games; but it has come to mean just that, especially in the extended, metaphoric sense in which we talk about "the game of life" or "the evolutionary survival-game." There is supposed to be something tough, self-mocking, or heroically defiant in talking about life in games terms and in being ashamed of the seriousness of words like "reality." To anyone who allows himself to get into this frame of mind, map making and its rules will also seem like one of the many "games" we play, but at least this is one game whose rules do not make any sense at all unless we constantly refer to "reality" and constantly maintain the distinction between the map and the reality-being-mapped. What is the scale? A proportionality between the area of the map and the area of the reality-being-mapped. What is distorted in systematic distortion? The shape of the reality-being-mapped. What is distorted in selective distortion? Elements of the reality-being-mapped calling for special attention. What is omitted in the omissions? Elements of the reality-being-mapped that are unimportant for the purposes of this map. The only thing that could be called "arbitrary" about a map, once its function has been decided, is the actual devices used in the symbolic content—whether dotted lines or cross-hatchings or colors or circles or squares shall be used to stand for this or that item in the reality-being-mapped. And *that* decision is not so much arbitrary as determined by the need for clarity in the picture or gestalt made by the map: if dotted lines are to be used for boundaries, they should not represent railroads, on the same map.

At the present time, the limitations of good science cannot be overcome by playing the acceptable language games of science better, but by getting better models of the phenomena under study and better control of the variables that the models involve. The mystery shows up in all that has to be omitted to increase the certainty of models. But a science from the middle section (Table 1), like psychology, ought to provide a field day for the linguistic analysts. Even in its simplest, most "mathematical" parts, psychology continually challenges us with questions like the following. What do intelligence tests test? The person tested or the designer of the test? What is the proper *usage* of a word like "intelligence" in science? In everyday life? In history? In philosophy? In art?

What is the future for Linguistic Philosophy, in the sense of a good, desirable, useful, hopeful future? Well, it is not in beating dead horses, refighting Napoleon's battles, and inventing more abstract rules for even more abstract talk-about-talk-about-talk. But must every candidate for this school recapitulate the philosophical biography of Wittgenstein in moving from Positivism (with its quest for the ideal language-game for science), through semantics (with its recognition of the functional basis of language-using), and ending with linguistic behaviorism? Yes, in imagination he must recapitulate it, if he would understand this piece of the history of philosophy. The *Philosophical Investigations* ends at the threshhold of linguistic behaviorism but does not step over it. Although richly larded with quasi-psychological and quasi-epistemological observations, it proposes no theories of either psychology or epistemology. Wittgenstein was too modest to make such claims, and he still believed that philosophy is not an empirical discipline.

109 . . . We must do away with all *explanation*, and description alone must take its place. And this description gets its light, that is to say, its purpose—from the philosophical problems. These are, of course, not empirical problems; they are solved, rather, by looking into the workings of our language, and that in such a way as to make us recognize those workings: *in despite of* an urge to misunderstand them. The problems are solved, not by giving new information, but by arranging what we have always known. Philosophy is a battle against the bewitchment of our intelligence by means of language. (3)

Consistent to the end, *Philosophical Investigations* promulgates no conclusions. But it does embody a certain attitude toward language, an attitude of resignation and acceptance before the curious linguistic behavior and misbehavior of mankind, an attitude almost religious in its humility and wonder, with not a little hint of *"tout comprendre, c'est tout pardonner."*

But now the historical situation has changed. Positivism is dead, not because the languages of the sciences were not good enough, but because the behavior of the scientists was not good enough. The original science-glorifying followers of Auguste Compte and the French "social physicists," who gave this movement its peculiarly affirmative name, had assumed that scientists would of course be paragons of rationality, objectivity, impartiality, and (illogically) unselfish devotion to the benefit of mankind, and they could point with pride to numerous individual scientists who were just that. How could anything bad come out of science? But then it turned out that there were "pure" scientists, who were driven "purely" by an insatiable curiosity and the need to make order out of chaos, and "applied" scientists, who wanted to apply particular portions of theories to the solution of practical problems, with obviously good or bad consequences to someone. Of the latter, we could conceivably ask that they take a broad form of the Hippocratic oath, although outside medicine, none of them did. But could we ask a "pure" scientist to take the Hippocratic oath—for example, the discoverer of a new branch of mathematics or astronomy who sought no practical applications whatever?

Came wartime, and all those scientists, pure and applied, who did not positively rush to supply the arsenals of their respective warlords, found themselves classified by their governments as a "natural resource" and put to work at the war-winning tasks. And the more the wars became worldwide, the more it became treasonable to be devoted to mankind as a whole. A physicist with so much as a qualm of conscience over the uses to which physics was being put was a bad security risk, shunned by his tribe, if not actually penalized.

Even in so-called peacetime, the new truth-seekers behaved no better, and no worse, than the most ordinary of the Ordinary Language users. They had to live, didn't they? Mostly, they classified their own brain power as a "commodity" and sold it to the highest bidders: in industry, in communications, in entertainment, in education, in space travel.

Until recently. After the American competitive urge to beat the Russians to the moon was temporarily satisfied, producing the boredom that generally follows satiation, American attention shifted to the question of what man is doing to spaceship Earth and how long it will be before he makes it unlivable for *Homo sapiens*, precisely by his aggressive, exploitative, competitive, and callous use of his *sapientia*. (This extended metaphor may turn out to be space travel's greatest contribution human survival.) The drop in enthusiasm for science after the moon walks and after the publicizing of ecology was as sudden as a frost in September, overnight in cultural terms. In academe there was a loss of students and a rise of unemployment in the exact sciences. The usual scapegoating in abstractions had begun—it was *science* that was to blame, and specifically the exact sciences that had "unleashed" the technological powers now poisoning the planet. Even in industry, a distinctly Luddite hostility against technocracy appeared. For the eggheads (technocrats), having made the noneggheads (body workers) potentially obsolete through automation, were now turning to them and saying: "You lucky, lucky obsolete workers! Never in history have you had it so good! Just relax, watch television, raise carrots, and be content to be consumers. Consume the goodies of technology at the rate we technocrats will feed them to you, in keeping with the gross national product, which we will increase by exhausting the world's natural resources in two generations." All parties to this American dream are nevertheless counting on the exact scientists to come up with new, clean, manageable, inexhaustible sources of power, plus immaculate contraptions for recycling used materials and disposing of waste.

At the present writing, in the terms of cultural fashion, the physical sciences are out of it, passé, in the doghouse, needing to redeem themselves. Astrology is in; it is the age of Aquarius; the pseudosciences are flourishing, from phrenology to Advanced Fortune Cookies. In academe, if there are any optimists left, in the old-fashioned Positivistic, confidence arousing sense, they must be in the social sciences, since that is where the former would-be exact scientists are gravitating. Far be it from me to belittle what has been done in mapping the difficult and treacherous terrain of the social sciences, but a warning is owed to the new young explorers, who are mostly idealistic, wanting to be helpful, and above all wanting to deal with people rather than with atoms. At least the foggier bogs and marshes should be posted, although they may be utterly fascinating, leading investigators unwittingly into literature: the marginally verifiable, the undisverifiable (the graveyard of abandoned theories), the strictly unrepeatable (all history), the as-yet inconclusive, the imponderable. Also, the obvious and the trivial. As David Riesman put it, all that research and all that foundation money, just to learn what de Tocqueville knew more than a hundred years ago! But they will certainly learn something. They will learn about the decisions that must be made, on what to put in and what to leave out. And how different it looks, from different angles of perspective. How many different angles of perspective are there?* But surely

* Regarding perspective, the cultural anthropologist Paul Radin (*Primitive Man as Philosopher, The Trickster*, etc.) used to regale us with stories about his profession. It takes only three things to "do" anthropology: one tape recorder, one Indian, and one piece of foundation money. Place the tape recorder in front of the Indian and start it. Ask a few stupid questions to start the Indian. Use the foundation money to publish what the Indian says, and you are in business. Of course, if you want a more well-rounded view, you will have to wait until the Indian places the tape recorder in front of the anthropologist, and so on.

there are some variables that are objectively measurable, such as crime and the death penalty? Is the electric chair, for example, a deterrent or is it not? Amazing, how difficult it is to establish a coefficient of correlation between only two, easily quantifiable variables—say, the consumption of tobacco and the incidence of lung cancer. Still more amazing, however, is the reaction of consumers and industry when "the facts" are made public. But we must keep trying. Amazing, how easy it is to get deluded into thinking that because we are dealing with numbers, we are approaching the status of a "hard" or "exact" science that deals with people. And so we have the pollsters, the survey makers, the statisticians, the systems analysts, each with something positive to contribute. After all, who would not want to be warned of an approaching catastrophe, the way the weather forecasters can warn us of the approaching hurricane? By all means, let us use these "tools of the social sciences"; but let us know of their reliability, and let us ask who is using them, for whose benefit, and at whose expense.

Is there such a thing as a "pure" social scientist, in analogy with the "pure" physicist, seeking no practical applications, but motivated purely by insatiable curiosity and the need to make order out of chaos? A philosophical question, an ethical question, basically, *ought* man to study his fellow creatures in social groups as if they were stars or galaxies? But there is a new breed of social planners, called futurologists, who sit fascinated before thousands upon thousands of "simulations" of everything that will be the case at t_1, t_2, t_3, . . . , if, if, if, if. . . . These worldwide hypothesizers are admittedly concerned with influencing worldwide changes, so one could surely be forgiven for asking who finally will decide which simulation is "the closest to reality"? And forgiven for observing how many opportunities for error and mistakes of quantification and omission there are, in all those flow-charts, all those feedback loops. Here we have the apotheosis of scientific humanism: tell us what your values are, and we will tell you how best to achieve them. Is your top priority peace? Good. Everybody wants peace. But if it is the military minds that are using systems analysis to get peace, they ask for "the worst case" and they assume we must prepare for that "at all costs." They recommend multiple overkill to the point of economic exhaustion, providing temporary employment for many. It was a five-star general, Dwight Eisenhower, who warned us of the takeover of American society by the military–industrial complex. Well, then, better the international-conglomerate complex? Or the third-world-developing complex? Are these futurologists willing to take the Hippocratic oath?

With that philosophical question we are brought back to Wittgenstein and the future of Linguistic Analysis. In performing our rough survey of the changes in the historical situation since Wittgenstein, we have touched on many new fields of opportunity for the exercise of Linguistic Analysis, and that in the precise Wittgensteinian sense of "doing battle against the bewitchment of our intelligence by language." Not just the bewitchment of the social scientists by their own jargons and their specialized "tools," but the far more widespread bewitchment that may result when their policies of public action become translated into the Ordinary Languages of voters, politicians, reformers, revolutionaries, slogan-makers, public-relations men, diplomats, and summiteers. What glorious opportunities for linguistic therapy! And who would not like to see a job of Linguistic Analysis done as a continuous crawl at the bottom of every television commercial? Linguistic pollution fills the air. Systematic lying-for-profit has turned even the 7-year-old language users into jaded sceptics, just as they are supposed to be starting out on their linguistic careers.

Table 1, which lists the range of mapping strategies and mapping devices used in the different kinds of knowledge, suggests where the future of Linguistic Analysis lies. Every one of these departments of knowledge needs a Linguistic Analyst in

Residence. Actually, each such department has many such analysts, but they call themselves critics, not philosophers. To criticize is to pass judgment on, to evaluate, to praise, to blame, to suggest improvements, to eliminate errors, to try out new angles of perspective, to clean up muddles. One cannot be a cartographer without studying his region, to see what has already been done, what ought to be done to improve what has already been done, to eliminate what has been done in justifiable ignorance or in unjustifiable delusion, to see what new, or old, or better devices can be invented or adapted for doing it. In other words, there is no such thing as "pure" description. There are only good or bad, better or worse descriptions. The battle cry of the phenomenologists "Back to the things themselves! Back to pure description, with the explanations held in suspense, in brackets!" is reminiscent of Wittgenstein's "no explanation, only description." But even the phenomenologists failed to realize that there is no pure description, only fuller or scantier, richer or poorer, description, always involving systematic distortion and selective distortion, drawing attention to or away from one thing at the expense of another.

As for systematic distortion, the "frozen philosophy" contained in the structure of each language which we called the map projection in Chapter Five, there is now a whole empirical science, structural linguistics, which tries to classify, compare, and conceptualize the great variety of structure among the 5000 known languages. As long as we are using a particular language, we are a party in agreement with the hidden philosophy that has structuralized the nonverbal reality in this particular way, a piece of cultural destiny as inescapable for children as their genetic heritage. But not entirely. Perhaps it is escape from this linguistic destiny that Wittgenstein had in mind when he said we must solve certain philosophical problems by "looking into the workings of our language, and that in such a way as to make us recognize those workings: *in despite of* an urge to misunderstand them."

I have already suggested that in view of the coming planetary organization of the world, there will be a great future for the translating business. Translating always involves a degree of philosophical Linguistic Analysis, even if the translator is doing it unconsciously. What is the equivalent of a Chinese expression in, say, Swahili? How do we translate intonation, gesture, the personal and impersonal connotations, the affective overtones, the allusions, the ritual and ceremonial meanings, the humorous, the insulting meanings? I now suggest that Linguistic Analysts enter the field cross-cultural communication, to try to reduce the sheer mass of world misunderstanding resulting from cultural provincialism and linguistic colonialism.

REFERENCES

1. Norman Malcom, *Ludwig Wittgenstein: A Memoir*, Oxford University Press, New York, 1948, p. 86.
2. Ludwig Wittgenstein, *Philosophical Investigations*, The Macmillan Company, New York, 1953, p. 4E. Used with permission of Basil Blackwell & Mott, Ltd.
3. Wittgenstein, *Philosophical Investigations*, p. 47E.

Nine Aesthetics and the Cognitive Theory of Art

Art begins where psychology leaves off, or, to stick to the verbal devices of our "bridge," poetry attempts knowledge of precisely that part of experience that psychology has had to omit in order to generalize and conceptualize from unique, individual cases. Donald A. Stauffer points up the difference in the Prologue to *The Nature of Poetry*, saying "this whole book rests on the assumption that a poem is like a person."

We do not enjoy a friend by reducing him to a set of categories: age, 30; hair, dark brown; nationality, U.S.; religion, Methodist. And the better we know the friend the less we are satisfied to ticket him as 30% justice; 30% cantankerousness; 20% amiability; 15% laziness; and 5% hot temper. Instead we take pleasure in recognizing that

> *I must confess it could not choose but be*
> *Profane to think thee anything but thee.*

A poem, just as clearly, is a fusion of innumerable elements, large and small, in many fields of experience, obvious or subtle or even not consciously noticed or describable. Experienced fully as a poem, it is not made up of separable or interchangeable parts, and its exactness lies in its unique wholeness. (1)

In aesthetics, the two most thoroughly cognitive theories of art I have been able to find are those of Susanne K. Langer and R. G. Collingwood, and here I must confine myself to the bearing of the present epistemology on their disagreements, as well as acknowledge my indebtedness to them both. They agree with each other that the object of art is the cognition of feeling through the creation of visible or audible forms, using imagination in the process, but they are not very united on how the feat is accomplished, nor on how imagination and feeling are related to "ordinary" life and to the constructions of science. According to Collingwood's theory of the imagination, language itself is originally expressive and imaginative in its function; but Langer is not willing to call the arts "another language" because the arts do not have a vocabulary, their "parts" do not have a conventional reference such as can be found for words in a dictionary. But this only means, in our terms, that the analogy must be made between the nonverbal arts and poetic language, not the arts and Ordinary Language, and least of all the arts and scientific language. The poetic language also lacks a vocabulary in the sense of the conventional references found in dictionaries; or it has one only in the minimal, starting-point sense that the poet must do his best to defeat, in order to avoid, the prosaic, and to make his poem truly untranslatable—a unique wholeness. Of course the language the poet uses is the repository of the categorizing and conceptualizing activity of his speech community, and he cannot entirely escape this; but it is a limitation of the poet's material that challenges his creativity, as the hardness of stone challenges the sculptor. In the repeated last line of "Stopping by Woods on a Snowy Evening" we can almost see Frost wrenching his more literal-minded readers away from the conventional, dictionary referents of "miles" and "sleep" in order to force them to try out the full imaginative possibilities of these ordinary words in this particular context:

> *The woods are lovely, dark and deep.*
> *But I have promises to keep,*
> *And miles to go before I sleep,*
> *And miles to go before I sleep.*

The advantage of stressing the analogy between the nonverbal arts and poetic language, as against Ordinary Language is, first, that we emphasize that in the arts it is the artist's imagination calling out to the appreciator's imagination, not just one classifying machine to another, and second, that thereby we do not raise the wrong expectations in the appreciator—namely, to find in the arts nothing but the psychologist's concepts of feelings, already the watered-down results of categorizing and generalizing from unique, individual cases.

But this does not mean that none of the psychologist's classes of feelings ever appear in the arts. The most primary and the most universal feelings appear in the most popular and universal arts: folk singing, folk dancing, decorative arts, legends and ballads, popular theater, puppet shows, and so on. Such feelings as love, hate, sadness, joy, yearning, grief, surprise, jealousy, and pride—are they not endlessly rehearsed in popular art, although even then with always a touch of individuality, or at least ethnic flavor? Think how the art of the mime is understood in all countries, or the art of the clown. Now if there are some forms of popular art we do not like, positively cannot stand, are bored to tears with, we should not blame the cognitive theory of art for that, because the same thing can be said about the feelings in the primary repertoire—they have been done to death by the repetitiousness of life itself. But even in the "fine" arts some of the simpler feelings appear—for instance, in the directions given to the performers in music or drama, or in the "high" moments of action. In music, how many different "sad" and "happy" movements are there in all the symphonies and sonatas written thus far? Obviously, it is not just for their sadness and happiness that we listen to these compositions, but for their unique differentness. It is as if, once a person learned the language of music, he could be carried by it on and out into realms and kingdoms of feeling-experience he never had in "ordinary" life and never suspected were possible. The same is true of the poetry language, the painting language, the sculpture language, and so on. If he can say in *words*, what these nonverbal art forms make him feel, he is a poet.

We have no choice but to live in the permanent mystery of the Heraclitean flux, but learn to ignore most of it through our practical and common-sense mappings of our everyday world. However, nowhere so much as in the arts is it brought home to us that selective distortion and omission are the price that human knowledge has to pay for clarification. Works of art are like teaspoonfuls dipped out of the Heraclitean flux and frozen into crystals. Thus they not only rescue moments of clarity from the transitoriness and triviality of actual experience, but, as is often forgotten, they would lose this value if the flux itself threatened to congeal into a glacier, as it does in static metaphysical viewpoints, material or spiritual. Selecting, shaping, leaving out—that is the price of achieving a gestalt, whether in the writing of a sentence, the carving of a stone, the making of a map, or the editing of a movie. In poetry, "too long a catalogue risks being more like a telephone directory than a poem. Do Whitman's longest geographical lists make America seem great, or merely overstuffed? . . . And is not Homer's catalogue of the Grecian ships itself less akin to poetry than it is to Burke's *Peerage*?" (2) The crowded, the cluttered, the chaotic, these tire us and bore us in art more than in everyday life, for we expect something different, and if we have to defend ourselves against *art* by inattention, what good is it?

Art is the embodiment or enactment or reenactment of selected and clarified life experiences in a limited sensory modality, and after the chapters on imagination and on the emotions, we can see a little better how the art materials themselves help, or force, the artist to select, shape, and leave out. Should all the arts aspire to the condition of music? Surely not. We know now from the tripartite characterization of the primary emotions that all the arts involve (1) the evoking situation, (2) the behavior in or toward the situation, and (3) the subjective feelings. We agree with Mrs. Langer that music is the tonal analog of the subjective feelings; but music tells us nothing about the evoking situation nor about anyone's behavior. That is what makes music so transferrable, as a component of the background in so many similar situations and behaviors. But why should this condition limit the other arts? Fiction, for example, is often content to describe only the situation, and/or only the behavior, leaving it to the reader's empathic imagining to supply the subjective feelings. Think how, without a word about feelings, Albert Camus builds up terror in the imaginative modern reader in the opening pages of *The Plague*, with his purely factual description of the ignored dead rat on the staircase. Moreover, consider how the device would not have worked if the author had not been able to count on the reader's knowing the connection between rats and bubonic plague. For the artist, the different art materials are like different vocabularies, in which he learns, from his own and others' trying, that some things can be said and some things cannot be said.

The chief difficulty of extending the mapping model to the arts is the difference between "embodiment" and "enactment or reenactment," that is, between the finished and the performative arts. For a map is a picture, not a set of actions. But a map is eminently the kind of picture that can be used as a set of directions for actions, where to go next, and what to avoid. In the case of the performing arts, the mapping model must be applied, not to the performance itself, but to the scenario, script, or musical score, which tells the performers how to reenact the author's intentions.

In working with Collingwood's distinction between magic-art, amusement-art, and art-proper, we are dealing with what art can be used for, rather than with what art is. That art can be used to arouse people to action or for entertainment should not surprise us, for the same can be said of language, and we would not call only its cognitive function language-proper and the other two magic-language and amusement-language. Of the three, magic-language would have to be the most primordial, since in their origins both language and gestures were intended to induce appropriate actions promoting tribal survival in specific situations. We are only now beginning to understand the sound-and-gesture languages of the other animals on this basis. But the magic-language would not have worked for survival if it had not contained at least some degree of correspondence with the important elements of the tribal survival situation.

What is lacking in Collingwood's theory of the imagination is that part which we have called "vain imaginations," although he seems to be groping toward something of the sort in his idea of the "corrupted consciousness." In an enterprise so dominated by the imaginations of both the artists and the appreciators, it should not surprise us that vain imaginations flourish most luxuriantly, not only in the uses to which art is put, but in the works of art themselves. Man's endless propensity toward lying, his fear of truth, and his self-glorification by means of self-delusion and fantasy—do these not meet us everywhere in the history of art, and especially in those monumental works which have survived to us from the self-glorification of powerful empires and power-wielding priesthoods? The trouble is that puritanical or iconoclastic revolts against the arts, brought on by excesses in the sin-revealingness of the arts, do not succeed in changing human nature, but only in hiding man's vain imaginations from himself and forcing the arts into less obvious, more "spiritual" forms, such as self-righteousness in place of self-glorification and hypocrisy in place of lying.

We should, however, especially heed Collingwood's warning that our culture has gone amusement-mad, because now there appears to be a real possibility that considerable portions of our population will be rendered unemployable through automation, and some of our social planners are proposing to fill the void in their lives with amusement-art, even though in large doses it leads to boredom and apathy. The catharsis principle does indeed rationalize the theatre arts, but now that these arts have been transmogrified into a mass-entertainment industry, they threaten to paralyze our citizenry in front of their television sets. However, Collingwood did not appreciate that even the greatest art, the most "cognitive" art-proper, can be used in this way, if the boredom is desperate enough. Yeats saw the situation more clearly. Asking to be delivered out of the flux of nature, "the mackerel crowded seas," into the province of art, the "artifice of eternity," he goes on to say:

> *Once out of nature I shall never take*
> *My bodily form from any natural thing,*
> *But such a form as Grecian goldsmiths make*
> *Of hammered gold and gold enamelling,*
> *To keep a drowsy Emperor awake;*
> *Or set upon a golden bough to sing*
> *To lords and ladies of Byzantium*
> *Of what is past, or passing, or to come. (3)*

Here is art-as-the-cure-for-boredom. Here also Yeats voices the aging man's Platonic yearning that underlies the formalist theory of art, which is not only to escape from the flux of nature by making finished, perfect things, but by becoming oneself a finished, perfect thing.

In the arts it is less easy to forget the permanent background of mystery than in the sciences, because the order made out of chaos is here so fragmentary, so dependent on the individual imagination and talents of the artist, so unpredictable, and so at the mercy of historical accidents and the freakishness of preservation and destruction. It is a greater temptation for the individual artist than for the appreciative public, especially in an age when artistic creativity has become the religion of many intellectuals. Whereas in the past the artist was an innocent, obedient craftsman, hired to promote the private or public idolatries of the rich and the powerful, ever since the Romantic period he has been encouraged to worship publicly at the shrine of his own godlike creativity, with the disastrous result that such a worshipper has less and less to say to anyone about anything, tending to talk more and more in public to himself about himself—or else to exteriorize every impulsive novelty he can think of as the latest revelation from the Divine. It may well be that we are at the end of an era in the arts, when even the pact with the Devil will not produce results, and utter exhaustion begins to show. More sober-minded artists are content to return to the status of craftsmen and to explore with craftsmanlike curiosity the artistic possibilities of new materials, including the new products of technology—and of course no one can predict the future there.

In the ancient and battered medium of words, however, the greater the artist, the more he takes truth as his criterion; and truth means acknowledging the mystery, not ignoring it, or denying it:

> The serious fiction writer will think that any story that can be entirely explained by the adequate motivation of the characters or by a believable imitation of a way of life or by a proper theology, will not be a large enough story for him to occupy himself with. This is not to say he doesn't have to be concerned with adequate motivation or accurate reference or a right theology; he does; but he has to be concerned with them only because the meaning of his story does not begin except at a depth where these things have been exhausted. The fiction writer presents mystery through manners, grace through nature, but when he finishes there always has to be left over that sense

of Mystery which cannot be accounted for by any human formula. (4)

We cannot here go into the several arts separately, although that is the proper way to discuss them, since they are quite different from one another and each has a unique history. We must end by touching on two factors that are more concerned with the art appreciator than with the artist himself. These are the "distance" factor and the "excitement" factor, two inevitable ingredients of art criticism.

Much has been written—much that is muddled and much that is confessional—about the famous "aesthetic emotion" or "aesthetic distance" that the art appreciator is supposed to feel while he is in the act of contemplating a work of art. The muddle, of course, arises from the failure to distinguish the feeling that is in the attitude of the appreciator *toward* the work of art and the feelings, usually too complex for words, that are cognized, or re-cognized *in* or *by means of* the work of art. But if truth, the truth of feeling, is the goal of art, rather than beauty or moral goodness, then the basic attitude of the art appreciator must be similar to that of the scientific truth-seeker, especially in the "pure" or nonapplied sciences such as astronomy or cosmology. It is a combination of curiosity and the need to find or create order out of chaos. Curiosity alone is not enough, for that is an animal feeling that is easily aroused and easily satisfied, and those who indulge it in aesthetics become novelty hunters, band-wagon jumpers, shock-appeal exploiters—anything, anything to avoid the boredom and nausea of satiated curiosity. Animals, when their curiosity is satisfied and their situation relatively secure, go to sleep. What is man to do with his curiosity when his situation becomes relatively secure? He contemplates the order that he finds already built into the universe and wonders what it means. Curiosity alone would never have given rise to science, just to old wives' tales and gossip. There has to be in addition the entirely human needing and finding and creating of order out of chaos.

The cognitive *attitude* toward the work of art releases us from at least some of the absurdities and irrelevancies that multiply in the wake of the other two attitudes—the erotic, whose object and goal in art is beauty, and the moral, whose object and goal in art is moral uplift or the emulation of an ideal. Both are real elements of life and are certainly plentifully represented in the *contents* of works of art. Here we are concerned with what happens when the erotic and the moral attitudes determine the attitude of the appreciator. First we discuss the erotic attitude. I pass as lightly as I can over the confessions of the art critics of the beauty school when they are talking about what art "does" for them and what it should do for the reader of the criticism. They are called *connoisseurs*, and reading any one of them we soon learn what it is he is so well informed about. It is rather like listening to a gourmet talking about his more delectable discoveries and favorite hates; there is much drooling and mouth-watering; there is yawning, scorning, rejecting and spitting out; there is cultivation of new variations of old favorites and exploration of rare, foreign, or outrageous treats. On a broader scale, soon the ugly is found to be more interesting than the obviously pretty, the evil is declared to be more exciting than the good, the tragic becomes "a difficult beauty," and in political violence and chaos "a terrible beauty is born." Since the art, to be approved, must be enjoyed, we are trapped.

Let us pause a moment (an hour would be better) before Picasso's *Guernica*, to see if it is appropriate that we should "enjoy" it. Its content—the animal and human pain and brokenness, the shrieking women and the dead children, the brute force and nightmarish light bulb, the feelings expressed *in* it, insofar as we reenact them in imagination just by seeing Picasso's selective distortions—is utterly repellent. And we are repelled; there is no beauty here, nothing to enjoy. Our approval of the painting consists of cognizing, recognizing, appreciating, that a good job has been done in characterizing the quality of the sufferings of war. It certainly does not mean that we approve of war or of suffering, or, as the Freudians would assure us, that in fantasy we are sadistically enjoying the inflicting of suffering, or masochistically enjoying the enduring of it. I repeat: the content is repellent; I would not use it to wallpaper my breakfast nook, and so far I have not noticed the army using it on recruiting posters. But it is a good painting, even a great one, from the cognitivist standpoint.

In a milder degree, all the other "negative" feelings expressed in art are cognized for what they are in their qualitative forms, not "enjoyed": funeral marches, tragedies, sad movements, melancholy poetry, memorials to the dead or to catastrophes. The cognitivist in art criticism is finished with his job when he has pointed out the elements that have gone into the selective distortion of the work of art and has shown how they react on one another in the finished work. To continue the cooking analogy, he is like the unscrambler of an omelet, pointing out the different eggs and other ingredients that went into the making of the dish but not presuming to say that the omelet is "nothing but" the eggs. Nor does he presume to say that people ought, or ought not, to eat only this or that kind of omelet. That is the educationalist's job—or the moralist's job or the religionist's job. Nor does the cognitivist presume to pry into the private life of the cook (artist) to see where he got the eggs. That is the biographer's job. I am afraid that the cognitivist attitude in art would generate much unemployment among art critics, especially those who write for the popular press, for it does not generate manifestoes, fights, personal assaults, excitement. It generates, rather, in the case of literature, a second order of literature about literature, often of higher qualitative content than the work of art being criticized.

Next, we deal with the moralist attitude. The excitement factor is the chief concern of the moral-uplift or ideal-emulation school of art criticism. We all start out as children, and children are first attracted to music and dancing by the excitement these activities create or "work up" in the participants: the faster beating of the heart, the faster breathing, the jumping, stamping, clapping, drum-beating, leaping, waving, rhythmic, repetitive motions and sounds organized by a leader. All these seem as "natural" in the kindergarten or playground as games or sports and hardly distinguishable from them. But we know now that they are as "super-natural" or beyond-animal as the long period of childhood dependency itself. Whether regarded by grownups as "recreational" (letting off "animal spirits") or "educational" (learning controlled action in a group), they are in both cases the child's introduction to the excitement of art. The excitement is real, not illusory, and requires healthy equipment for its enjoyment: good muscles, bones, glands, circulation, sensory input and physiological output, and some degree of mental control or direction. Here again we graze the question of how much of this excitement is made possible by the genes, how much by economic supplies, how much by individual temperament and cultural encouragement. The lucky ones continue to increase their enjoyment of the excitement in proportion to their health and nutrition as they get into their subteens and teens. What a tribute to the luxury aspect of Western technocratic civilization are all those screaming, clapping, gyrating devotees of song heroes and band groups and buyers of their records. The promoters make money and step up the excitement, which makes more money and steps up more excitement—another "benign" economic cycle that keeps a great many unemployed off the streets. Why should we be surprised at their Dionysiac frenzies, their communal solidarity, their in-group speech and dress? Teen-agers, whose physiological equipment is at its prime, want to *be* excited, to feel the excitement in their *guts*, to feel alive and free and active, just at the time when the elders want to cast them into pre-ordained social roles and to saddle them with burdens, weights, obligations.

But we have already seen in Chapter Six how the long and expensive process of human growing up differs from the animal growing up, and how the parents must use the animal emotion repertoire to integrate the child into the chosen or inherited adult value system. And they certainly use the excitement factor as soon as possible. They teach their children "the legends of our people" in terms of the celebration of anniversaries, holy days, and festivals. Parades, military marches, dramatic reenactments, memorial pageants, folk songs and anthems, all try to create excitement and move the participants to exaltation, yearning, nostalgia, pride, dedication, willingness to sacrifice, feeling part of something greater than oneself. And we feel the excitement, even when we are partially alienated from the culture in question, or hostile to it. In the last case we say the people are being brainwashed, manipulated, made to dance like puppets on a string. But are we not all brainwashed, manipulated, made to dance like puppets on a string since earliest childhood? And are we not all *conscious* of it precisely to the degree that we are able to stand outside it, in imagination watch it being done, in freedom giving or withholding our response? But that is growing up. If and when our own views were to prevail, we would have to resort to the same tactics, write a new anthem, start a new parade, create or join a new ritual community of shared allegiances. (Even Kierkegaard, who said "the crowd is untruth" and recommended that the truth-seeking individual stand alone before God, did not keep his mouth *shut*. He wrote for other lonely individuals standing before God.) As we withdraw our imaginative sight from vertiginous views of the order in the stars to even more vertiginous views of the order of life and history here on this planet (that local enclave of antichaos), we know that we stand both inside and outside an order of some sort, in some stage of its development. As Eric Voegelin puts it, the order of history emerges from the history of order. But what is the order, and what stage of it are we in? Are we still children, spiritually, no further developed than the Bronze Age in the feeling *contents* of our cultural treasures? Or are we senescent, grasping after outworn creeds and metaphors in our elderly efforts to clutch for security and comfort?

For the near future the culture tasks seem increasingly obvious. The entire human race must be gradually transmuted into a single tribe, but certainly not a tribe of the Bronze Age pattern. Multilingualism must be started at the earliest possible stage. Some code of international good manners and bad manners must be taught to the young as we now teach toilet training and table manners. The entire educational system must be rearranged according to what is most beneficial to which age group, and it must become a life-long process. Each individual must be encouraged to recapitulate the entire history of mankind in some ritualized form, as far as his genes permit. Genetic differences must be recognized. The noninheritance of purely learned characteristics must also be recognized. National, cultural and racial differences must be appreciated, not homogenized; but it is necessary to deprive them of their political significance. Politics must become housekeeping, both on the regional and on the worldwide scales. Sports must become the moral equivalent of war, again ritualized and controlled, as an appropriate outlet for those Bronze Age attitudes which we have outgrown and no longer admire in other areas of life—too much aggressiveness, competitiveness, egomania, and callousness. At least let us keep those sport heroes out of politics and out of the military. As for nonfunctioning tribal hierarchies of governments, churches, universities—let us not waste emotional energies fighting them. Let us just withdraw our support and admiration and let them quietly dig their own graves. There is too much work to be done elsewhere. And no time for boredom.

As for the far future, who can tell? Will it be the greenhouse effect or another Ice Age? The last Ice Age in Europe, some think, forced the larger animals in the north southward, trapping Cro-Magnon between the caves and the Mediterranean, and changing his character from the milder manner of the tropics to the fierce hunter who could outwit, gang up on, and triumph over, predators which were five, ten times his size. Is that where it all started? But the Bronze Age did not come until agriculture and property became possible and had to be defended and fought over. What a pity: the entire human enterprise lasting only from one Ice Age to the next, when the other animals have done so much better. But if man does survive in some form, how can we tell now what fugues, what dances, what anthems will warm his heart and stir his mind? Will he look back at us now as we look back on Stonehenge? He will probably have our literature, but what metaphors will appeal to his imagination, knowing of the monumentalities of Boston, Paris, Moscow, and Peking crumbling under 5000 feet of snow and destined to stay there tens of thousands of years? But let us not expend our emotional capitalon these vain imaginations—"O the mind, mind has mountains, cliffs of fall . . . ." Rather, let us turn to the tasks at hand, the tasks of the near future listed in the previous paragraph.

REFERENCES

1. Donald A. Stauffer, *The Nature of Poetry,* W. W. Norton & Company, Inc., New York, 1946, p. 16.
2. Stauffer, *The Nature of Poetry,* p. 62.
3. Reprinted with permission of the Macmillan Company, from *Collected Poems* by William Butler Yeats. Copyright 1928 by the Macmillan Company, renewed 1956 by Georgie Yeats. With permission of M. B. Yeats and the Macmillan Company of London and Macmillan Canada. Stanza IV of "Sailing to Byzantium," p. 192.
4. Flannery O'Connor, *Everything That Rises Must Converge,* Noonday Press; © 1956 by Farrar Straus & Giroux, Inc., New York; Introduction, pp. xxvii–xxviii.

Ten Psychoanalysis and Psychotherapy

With psychoanalysis and psychotherapy we are back in the middle section of Table 1—the section we characterized, to specify its differences from both good science and good poetry, as various mixtures of mediocre science and mediocre poetry. When we think of the many different aspects of man that psychology is supposed to cover, we see that it is an excellent example of a "soft" science. It can never become good science—there are too many variables, not enough controls, and too many imponderables and unrepeatables—but neither can it become good poetry, because it must generalize and conceptualize from unique, individual cases, and it must after all try to construct theories that are to some degree, no matter how slight, empirically verifiable.

In addition to being "soft" sciences, psychoanalysis and psychotherapy are normative studies about halfway between medicine and ethics, since they are concerned with applying the norms of mental health and mental sickness, as against physical health and sickness for medicine and right and wrong conduct for ethics. It makes little difference if, for ethics, we substitute culturally desirable and culturally intolerable behavior; they are just as normative. The proliferation of Freud's theory into the multiplicity of "schools" we have today should not be surprising, but neither should we attribute it to the youthfulness of this young science, as if we expected more scientific procedures in the future (i.e., a "hard" science). All the soft sciences are characterized by the gradual accumulation of good and bad hypotheses, because it is impossible to devise a crucial experiment to decide between any two of them. Where would chemistry be today, if the neophlogistonists were still in the running? But think of the accumulation of "cures" in medicine that cannot be proved or disproved. Such limitations of human knowledge we had better not disguise to ourselves, and yet it is no excuse for being less empirical than the subject matter will allow.

When we look back on Freud's theory from the standpoint of the evolutionary theory of emotions presented in Chapter Four, what strikes us most is Freud's utterly simplistic and mechanical understanding of the animals—no doubt the "scientific" opinion of his day. The libido of *animals* was thought to be an undifferentiated, blind, instinctual "drive," like a steam-propelled engine, for keeping the species alive. In man, this drive had to be repressed for the sake of civilization, giving rise to the neurotic development. Freud's virtual identification of the libido with sex makes it even more difficult for us to understand the higher animals, because we know that they mostly have their sex on a seasonal basis and if their *survival as individuals* depended on sex, they would die between one mating season and the next. And finally, Freud's inability to account for aggression in the unrepressed, sex-driven animals made him invent the death instinct, a "blind" provocation to violence to bring about the provocator's own demise. Needless to say, from the evolutionary standpoint, *there is no death instinct*. Death is very well taken care of by predation, starvation, disease, and environmental change. Not even the few animals who kill themselves when attacked, such as wasps, go about provoking such attack. Of course human beings, being capable of planned, rational suicide, are also capable of death-inviting behavior; but if any instinct is involved, it is one half of the pleasure principle, namely, to reduce pain. Suicidal drug users, the insane, prisoners, for example, are trying to get rid of an intolerable state of consciousness. On the other hand, death-defying danger seekers are trying to enjoy the thrills of narrow escape. A little reflection will show that a death instinct and survival of the fittest cannot be put into the same bag of explanations for evolution.

This biological by-product of Freud's theory—that it makes even the behavior of the animals hard to understand—should be a two-edged warning to us. Not only must we know a great deal more about the animals before we can use them to illuminate the animal continuity of our inheritance, but we must be extremely cautious in making comparisons with animals, assuring that we do not deny our discontinuity with them. We found the three most inescapable forms of our discontinuity with the other animals to be first, the oversized brain with its early-stage inefficiency; second, the long period of childhood helplessness, which is the price that must be paid for the maturation of the oversized brain; and third, the societal factor, which is the survival solution of this perculiarly handicapped species and which makes all sorts of beyond-animal demands on human beings, in the most primitive human societies, let alone the ones we now call civilized. In all strictness, we should compare modern man only with Stone Age man, not even with present-day primates or other higher animals (which have been evolving all along in relation to their environmental niches, to make their primary-emotion patterns all the more subspecies-specific). It is only with Stone Age man that *cultural selection* unmistakably adds itself to, or interferes with, natural selection. And it is only with Stone Age man that we begin to see unmistakable signs of such beyond-animal feelings as individual self-consciousness, other-selves-consciousness, past-and-future consciousness, tribal consciousness, death consciousness, world consciousness, and beyond-world or totality consciousness. Finally, it is only with

Stone Age man that we have unmistakable signs of the *ordering of values*, that social and individual subordination and superordination of passing emotions that brings about a degree of freedom from them and a unity of the "soul" in relation to long-range goals. Even in the most primitive tribes we know today, there is no indulging in sex while preparing for war. (Women are regarded as unclean and dangerous for this reason in the Stone Age tribes of New Guinea.) I particularly stress the dangers of comparing modern man with any animals other than his Neolithic forbears because the many fascinating studies of animal behavior now being produced by ethologists are flooding the popular market with grist for the "nothing but" syndrome. We should also stop attributing to the animals various human traits such as cruelty, malice, and revenge, and all positive or negative feelings that would be impossible in the *absence* of the above-mentioned states of consciousness.

But another warning is needed. It will no doubt be objected by some that the understanding of the animals provided by the evolutionary emotion theory is "anthropormorphic." We have, as it were, to humanize the animals in order to see any analogies at all between their behavior-toward-a-situation accompanied by empathically imagined subjective feelings and the human introspective feeling-reports in various human situations. But by now we should not use anthropomorphism as a term of reproach, for according to the epistemology presented in this book, our understanding of the atoms and the stars is also anthropomorphic—it is the outcome of the specifically anthropoid way of selecting, abstracting, categorizing, and model making in these regions of our experience. What is regrettable (or misleading, or useless) is misapplied anthropomorphism, as in astrology or in talk about "free will" in the atom. But psychology, of all would-be sciences, should try to be as anthropomorphic as possible, in the good sense of conformed-to-the-human. For if there is something wrong with "animistic physics," is there not something equally wrong with mechanomorphic psychology? We keep forgetting that it is man who invents life-imitating machines; machines did not invent man. Man's brain is not a computer—the computer imitates and enormously speeds up certain mental operations, but the computer is no more human than an adding machine or a wheel. It is an extension of the nervous system, as are telescopes, microscopes, and telephones; and like them, it is not "bugged" by the problems of consciousness, even the simplest pleasures and pains. Yet today we hear some scientists refer to their own children as biological computers programmed by heredity to react this way and that way. Garbage in, garbage out, including the garbage that man is "nothing but" a computer! What machine will it be next? Could this "nothing but" syndrome be the lazy man's way to satisfy his totality consciousness? Lévi-Strauss remarked that the *primitive* mind totalizes.

With all these warnings in mind, let us stand back and take the broadest possible view of the implications of the evolutionary primary emotion theory and the imagination theory, for the question of what it means to be "human" and what it means to be "normal." What we would all like to know, whether we are parents, teachers, or psychotherapists, is the rock-bottom inherited "genotype" or gene pool of humanity that, complete with individual variations, is partly born new and untouched into the world with each new generation and begins its interaction with the world from the instant of birth. The farther back we can put the line of demarcation between the animal and the human, the more universal will the traits be, and the less we will be misled by local cultural ideas about what is "normal" and what is culturally desirable, as Freud was misled by Victorian sex repression. Neolithic man was elected for being well over the line, but of course as soon as someone digs up the evidence for a much earlier culture, the "line" (half-a-million years wide) will have to be pushed back. The 20 to 30 thousand years between us and the last interglacial period are like the blink of an eye in the genotype variation time scale, although no doubt the rise of civilizations has permitted the survival of some phenotypes that never would have made it in the Upper Pleistocene (including your author). Of course, we could also argue the other way, that the rise of huge, oppressive civilizations obliterated some phenotypes that would have done splendidly in the caves. Think of the Egyptians, using men like sand to build the pyramids, and think of all the rigid hierarchies of history. All caste systems, being based on belief in the inheritance of acquired characteristics, tend to become self-fulfilling prophecies because of the role of expectation in character formation from the instant of birth. One who at birth is expected to become a prince is not going to develop the character of an untouchable, no matter what his phenotype, *and vice versa*. At any rate, here we are drawing the line prior to all the present-day racial and linguistic divisions of mankind, because they all have the culture-making characteristics.

Now these culture-making characteristics, in the form of the types of consciousness we have mentioned—do they not all involve the imagination? From the making of tools to the making of hunting expeditions, the building of campsites, the burial of the dead, the construction of mythologies and magic rites, dancing, celebrating—all involve the forms of imagination we delineated in Chapter Three, just to enable the ordering of actions in a means-to-end sequence and the necessary repression of some emotions and the encouragement of others around the desired goal. Here is the origin of the *ego*, in this emotion-ordering process. Yet animals have no need of it, since their goal is served by the unordered emotions as such, with some learning in each emotion dimension. And here is the origin of *freedom*, in the conscious repression of some emotions and the conscious arousal and expression of other emotions directed toward a conscious goal, usually socially defined.

This brings us to the ethical side of the psychotherapist's

job, which he can no more escape than can the doctor, on the medical side, or the parents and teachers, on the societal side. Ethics—from *ethos*, the character, sentiment, or disposition of a community—are learned first in the nuclear family. Not only the first "good things" and the first "bad things" are encountered there, but also the first "conflicts of interest." And not only sibling rivalry but every permutation and combination of father-mother-brother-sister rivalry. Every perceptive parent knows what all the great spritual leaders of mankind have known, but have explained differently, as the basic ethical problem of man: that every individual is born selfish and *must remain selfish up to a point just to survive*. Sometimes however, the individual continues to be selfish far beyond the degree necessary for survival, and far beyond the point conducive to the good of the family as a whole or of society as a whole. All over the world the rich get richer at the expense of the poor, the powerful get more powerful at the expense of the powerless, the healthy get healthier at the expense of the sick, and so on (make your own list). Whether we follow the Bible and call this phenomenon Original Sin, whether we call it Maya, the illusionary character of the world from which we must extricate ourselves, the phenomenon is there. Indeed, it is universal; and it, as well as the "solutions" to it, begins right in the family, the first society. Adam after the knowledge of good and evil can become Cain for no survival reasons, and refuse to be his brother's keeper. But he need not. He can also become "like unto God," like a good father. First love and first rivalry and yes, first self-transcendence, begin together in the family, as the parents gradually wean and woo, persuade and cajole, threaten and bribe, the innocent little egotists into responsible adulthood.

Responsible first to the family, then to the tribe or society, and then to . . . ? In the old days when the universe was small it was easy enough to put God in the final place, the place whose "referent" is man's totality consciousness, whether his culture calls it the All, the Ocean of Being, the Great Spirit, the Alpha and Omega, the Nameless, or you name it. Quite rightly the modern psychotherapist objects to being pushed into the role of philosopher-priest-shaman-guru to his patient. Since, however, in the process of effecting a cure there will have to be *some* reordering of values just to adjudicate the most glaring conflicts, what is he to do with the patient's own totality consciousness and whatever inchoate beliefs he may have about it? And what could "normality" mean, or "integration of personality," if the claims of all the beyond-animal forms of consciousness are not included and if the *fact* of their being at loggerheads, or split off into compartments while "residing" in the same brain, is not somehow taken into account? Surely psychotherapy cannot be more simple than human nature itself, when it tries to soothe us with "mechanisms" and "complexes" to blame things on and drugs to make consciousness bearable? Responsible psychotherapists of every school try to set the patient free *from the past*,—from whatever in his childhood or family situation is holding him in a vicious circle of rigid, defeating behavior—and to set him free *for the present and future*, for growing and learning from experience in some hopeful direction. Does that sound familiar? (See crude map of human consciousness, Figure 1*a* of Chapter Six.) Naturally this helpful intention is not promoted by appealing to the mythology of fatalism, such as the Oedipus legend, whose subject matter is neither incest nor parricide but the inexorability of fate: a man learns that a terrible fate is in store for him and decides to resist it; and then, by every step he takes to oppose it, he helps to bring it about. There is fate, both in heredity and in the family and historical situation, but there is also freedom; and, as Kierkegaard put it, the task for man's freedom is *to turn fate into providence*. Freedom-in-the-abstract is a delusion. Concrete freedom needs something to work on, namely, fate.

Redeem the past. Turn fate into providence. Do the best you can with what you have been given. The particular children that are born to parents are just as much fate to the parents as the parents are to their children. With increased knowledge, the role of responsible parenthood promises, or threatens, to become more difficult and crucial, especially as the societal demand for education tends to stretch out, rather than diminish, the long period of dependency. Today, "civilized" twentieth-century parents are being asked to change their child from a Stone Age baby into an educated, responsible citizen of an increasingly worldwide community, and to do this before the child loses all the vitality and flexibility of his youth and is well into middle age—surely a daunting task. It is easy enough to prescribe intelligent, foresightful love as the *sine qua non* for the parental role—as against stupid or selfish love. But it is precisely intelligent, foresightful parental love that demands knowledge to work with, in view of the harm that can be done through ignorance from the earliest stage onward. Perhaps the day will come when society will have to demand through education certain minimal requirements of natural parents, such as it now demands of adoptive parents, just to keep from increasing the populations of jails and asylums.

It is only to be expected that the primary emotion theory as modified by the imagination theory would turn out to be most helpful at the childhood stage of the life cycle, and possibly part of the way into the emotionally loaded teen-age period. For knowledge of the feelings of young adults, grownups, and older people, there is literally nowhere to go except literature (i.e., good literature; which is good precisely because it tries to do justice to the enormous complexity, concreteness, and uniqueness of grownup human "characters" and the excruciatingly complex human interrelationships that they manage to get themselves into).

Nevertheless, two hints and warnings can be derived from this theory for the problem of what it means to be normal. The first concerns the intensity range of the emotions in relation to their preservative function for the individual: in fact, neither at the lowest nor at the highest intensity can they perform their job of "intuitive appraisal of a situation as harmful or beneficial." This means that we should be very suspicious of any concept of normality that proposes to solve the problem of man's emotions by either of the extremes, either one that recommends the deadness of apathy or one that prescribes the blindness of the violent passions as ideals or paradigms of the normal human. We are tempted to choose one of the extremes because emotional control or self-management is so much a part of growing up and living in society that it seems that the task would be very much easier if we could either stamp out emotions entirely or give way to them unrestrainedly. But control is precisely for the purpose of keeping the emotions at a level where we can recognize them and learn what they are trying to tell us; not to get rid of them in order to feel nothing, not to wallow in them in order to enjoy their "feel." For the latter we should go to bad movies, which specialize in emotional clichés. Perhaps someday this emotion theory might even be taught to schoolchildren, to help them in their emotional self-management. We already know, mostly, what to do with simple, realistic fear: I spot the tornado cloud, feel the pang of fear, which I immediately repress to well below panic level in order to give my undivided attention to the learned protective behavior, but not so completely that I feel no fear at all and thereby get distracted in other directions or fall asleep. Finally, I feel the relief from fear after the all-clear is sounded.

The second hint and warning from the primary emotion theory for the concept of normality arises from the observation made in Chapter Six that

> Those of the simpler emotions which are in the medium intensity range, and are, therefore, not too weak to be ineffectual toward action and yet not too strong as to be impossible to maintain, physiologically, for more than a few minutes . . . tend to settle down into long-range attitudes, to crystallize as prevailing *emotion habits,* and

eventually to determine the overall character of the person as traits, dispositions, and temperament.

At the early childhood stage, before experience and imagination have had a chance to modify them, the eight primary emotions themselves constitute a minimum standard of normality (i.e., we would suspect something is abnormal if, under the appropriate evocative situation, a child of 18 months showed absolutely no signs of one or more of the primaries: anger, joy, acceptance, surprise, fear, sorrow, disgust, expectancy). But even a three-year-old is already so complex that he begins to show the signs of the first set of dyads, or normal, everyday emotions: pride, love (friendliness), curiosity, alarm (awe), guilt, misery, cynicism, aggression. Even on this level, which is simple enough, but which some "grownups" never get beyond, we must carefully avoid the Freudian tactic of explaining any one emotion as a perversion or substitute-gratification for love (eros); for example, interpreting curiosity, the motivation behind all the sciences, as a "sublimation" of sex. Rather, knowing the survival value of curiosity for the unrepressed animals, we should examine our childraising practices to determine how much they repress curiosity in the child, not for *his* survival but purely for the parents' convenience. Yet even on the preschool level an outside observer can begin to see differences of character or temperament between different children in the same family, and of course the mother can see them much sooner. And the teacher of the early grades can see that, under the same situation, certain children are *prevailingly* more fearful, more resentful, more accepting, more friendly, more disgusted, more curious, than others.

What are we seeing here? Differences in heredity? Just how early does the formation of emotion habits begin? And is it in imitation of parents, in reaction against them, or strictly individual, original? We all know the standard "national" traits exploited in humor and parody. An Englishman thinks the Italians are overacting and exaggerating their feelings. An Italian thinks the English are cold and clammy, an unfeeling race. It may be that as the situations the child must confront daily become more and more complex, it is more difficult to "appraise them intuitively" as being either harmful or beneficial or calling for a specific response, and there develops a tendency to *simplify complexity* by responding in a habituated manner. We have only to think of such a universal phenomenon as racial prejudice in terms of emotion habits. Racial prejudice consists of our preclassification of huge numbers of persons by some obvious differences from us, such as skin color or nose shape, followed by our habitual negative reaction to them as stereotyped class members, instead of taking the trouble to react to them as the different individuals they really are. Wrong? But it does simplify our lives, especially when our culture tells us in advance what our reaction to these huge classes of people should be. Or think how political rhetoric tries to simplify our lives, counting on our emotion habits in presenting the opposition in pejorative abstractions (and itself in euphemistic lies we most want to hear).

But let us not despair. Despair is the rhetoric of disappointed idealism, for which the cure is a return to realism, the realism of the biblical view of man. We have one great advantage over our ancestors: we no longer believe in the biological inheritance of cultural characteristics. Thus each new generation represents a chance for a new beginning, a chance to slough off the worst cultural accretions of the past. But neither do we ignore heredity: if something like musical genius "runs in the family," it is biological; the manner in which it is encouraged or discouraged or exploited is cultural. Geniuses are not madmen who turn their madness into works of art. They are persons biologically disproportionate in inherited capacities; and the disproportion tends to push them toward the psychically abnormal, whether in science or in art. Today we deprive geniuses of this romantic aura; we give them special schooling and let the less talented play the "role" of mad artist or mad scientist, if they wish. We recognize individual differences in children and the *fact* that these are, or can be, influenced by parental and societal attitudes and actions. But how? How much? In what direction? At what cost?

Concerning grownups, the situation is different. We simply confess our inability, at this point, to separate the inherited and the cultural factors in most adult human characteristics, however glibly we may categorize them in standard psychological pigeonholes. What are we to say of such characteristics of grownup modern man as his aggressiveness, his competitiveness, his egomania, and his callousness? How much is inherited, how much culturally promoted? And their opposites, passiveness, cooperativeness, humility, and sensitivity? Surely no civilization can survive without some or all of these characteristics doing their proper jobs in the right places. But it is easy to see that in the coming world-village, the first four are not going to be as useful as they were in the expanding, exploring, exploitative stage of the human species. And it is still easier to see how the first four, armed with atomic power, could exterminate the species. In case any women are reading this who might be inclined to regard these characteristics as masculine, let them recall that as soon as women were placed in positions of power by cultural history, they showed themselves expert at all of them, in their feminine versions. It was a queen who said: Let them eat cake! In our time, many virile males are getting fed up with having these characteristics culturally touted as signs of virility. Comedians, of course, have been making a living for years by joking about aggressiveness, competitiveness, egomania, and callousness.

To these four characteristics, callousness is the key, for if the callousness can be overcome, the other three will be certainly mitigated, if not dissolved or redirected to more socially helpful ends. The Bible includes callousness under "blindness," literally, *not seeing* the other persons as persons, but as objects variously categorized, nowadays, for instance, as the enemy, the competition, the consumers, the masses. And what other way is there to overcome callousness and to expand the egocentric human consciousness with the consciousness of others as persons, and with the effects of one's own attitudes and actions on other persons, except with the deliberate practise of empathic imagining and imaging-as-if-perceiving from the other person's point of view? In psychotherapy, the Carl Rogers school is based almost entirely on developing these abilities in counselors; the existentialist school tries to see what the world looks like to the patient, using Heidegger's *dasein-analyse* as guidelines to the human as over against the repressed-animal view; and practitioners of other techniques, including group therapy, use them because they work—or rather, without them, not much else works. Thus between this patient and this therapist, *something must happen that should have happened* between this person and his assigned human community of parents, relatives, friends, co-workers, the "world," and so on. And the therapist must not only help this particular patient "work through" his particular difficulties, he must give the patient the strength, courage, and desire to take up a constructive, realistic struggle with the social reality. What social reality? The same collectivity of selfish, rejecting, defensive, demanding, domineering, objects in collision, worlds in collision, which is the modern social horror, especially in big cities? Precisely. So let us not expect miracles from psychotherapy, especially when we ourselves are contributors to this unpropitious social reality, by our own inward "scale of values" or "order of priorities." Let us rather admire people for surviving at all, and for being counted as sane, normal. It means there are still pockets of redemption in the world, places where the forces of intelligent love are secretly at work.

Empathic imagining and imaging-as-if-perceiving from the other's point of view are not easy. When we achieve them with any degree of accuracy, we discover not only that people are similar, but that they are different; that the deep, individual, private, mysterious, uniqueness of subjective reality passes through the categories of psychology like water through a sieve. And that is as it should be, to protect the interior freedom and privacy of thought and feeling from would-be categorizers and brainwashers. Whatever people tell me I am, I do not have to believe it, and even if I do, I know that I am more than that, beyond that. And what are categories, words, but attempts to slice up the immensity of my concrete consciousness of totality as it flows through me and around me, into manageable pieces for public exchange? Only good literature gives us glimpses of the inward reality, and only for fictitious persons, because the privacy of real persons must be respected.

Eleven Heidegger's Ontology

Since there is such a yearning for some form of Bigthink in the world today, both to overcome the repressed-rat anthropology of the "nothing-but" psychologists and to fill the vacuum left by the linguistic philosophers, we must look at the most likely candidate for this double job, the ontology of Martin Heidegger. Does his ontology actually "correct" the "wrong turning" taken by ontological thinking in the early history of Western philosophical thought, thus constituting an improvement over Aristotle? In Chapter Seven we went to considerable pains to legitimize ontology (metaphysics) when this is undertaken as an enterprise of totality mapping, with all the difficulties of incompleteness, uncertainty, change, inclusion, selection, and omission that such a project would involve. But certainly such is not Heidegger's understanding of ontology, either his own or Aristotle's. He still believes that there can be a science of being *qua* being, which does not break down into the several sciences of different *kinds* of being.

R. G. Collingwood has stated with admirable clarity the case against the "science of pure being":

> The universal of pure being represents the limiting case of the abstractive process. Now even if all science is abstractive, it does not follow that science will still be possible when abstraction has been

pushed home to the limiting case. Abstraction means taking out. But science investigates not what is taken out, but what is left in. To push abstraction to the limiting case is to take out everything; and when everything is taken out there is nothing for science to investigate. You may call this nothing by what name you like—pure being, or God, or anything else—but it remains nothing, and contains no peculiarities for science to examine. (1)

But, of course, Heidegger does not admit this objection. Right at the beginning of *Being and Time* (1926), where he sets out his method, he *almost* sees it: "The Being of entities 'is' not itself an entity." But right there, in order to escape "nothing" to talk about, he must reify and substantialize "being" even as did Aristotle, and talk about it as a *something* that *belongs* to entities. Furthermore, in order to find the meaning of Being in general (the avowed purpose of *Being and Time*) he tells us we must give priority to the Being of that entity which inquires about Being, which we ourselves are:

Thus to work out the question of Being adequately, we must make an entity—the inquirer—transparent in his own Being. The very asking of this question is an entity's mode of *Being*; and as such it gets its essential character from what is inquired about, namely, Being. This entity which each of us is himself and which includes inquiring as one of the possibilities of its Being, we shall denote by the term "Dasein." If we are to formulate our question explicitly and transparently, we must first give a proper explication of an entity (Dasein), with regard to its Being. (2)

By this strategy, Heidegger temporarily escapes from the vacuousness of the concept of Being and gives himself something definite to talk about, namely, the human *kind* of being. But just as Aristotle, by talking about substances and causations turned his ontology into a kind of *Ur*-physics, so Heidegger by talking about the human kind of being turns his ontology into an *Ur*-anthroplogy, on even an ontologically verbalized psychology, however much he himself may deny it.

It was the psychotherapists who first made use of Heidegger's work and introduced him to the English-speaking world as a way of overcoming the repressed-animal anthropology of Freud, and surely this lends support to the suggestion that it is as a philosophical anthropologist, not as an ontologist, that they understand him.

The secret of Heidegger's "method" lies in the peculiar word "transparent" in the foregoing quotation. To "make transparent" is to look at the phenomena, and then to dismiss as "merely ontical" or "merely empirical" what the phenomena show, in order to "see through" them to what is "more primordial" and is, therefore, deserving of the appellation "ontological." The "more primordial" is to be thought of as that which is somehow "prior to" and also "makes possible" that which appears in the "merely ontical" or "merely empirical." For this writer, who has been using "empirical" all along in the widest sense of "experiential," it is difficult to see how the *results* of such ontological x-ray vision, assuming one had it, could be revealed anywhere but in experience—which would of course demote them once again to the "merely empirical" (experiential). Or is thinking *outside* experience? Heidegger repeatedly devaluates "the empirical" as if it were all on an equal plane of childishness or superficiality, whereas he ought to have known from literature that a good description of "experience" always contains many levels simultaneously—sensory, emotional, conceptual, imaginative, and so on.

On the other hand, Heidegger also insists that every Dasein "already" has some understanding of its own Being, since it is part of its nature to "comport itself" toward its own Being. "But in that case the question of Being is nothing other than the radicalization of an essential tendency-of-Being which belongs to Dasein itself—the preontological understanding of Being." (3) Ah, but is that not the whole point? At what stage, or at what age, in Dasein's "natural" growth in self-understanding are we going to start calling it ontological, at what stage merely preontological? Heidegger's criteria for deciding that remain to me unclear, opaque.

An example of this difficulty with Heidegger's "method" of arriving at "fundamental ontology" is his treatment of fear and anxiety. Following Kierkegaard quite closely at the start, he distinguishes the two by saying that fear is directed toward a specific danger from a specific quarter, and anxiety is indefinite, its object being "nothing" and "nowhere." But whereas Kierkegaard made it clear that "nothing" and "nowhere" are applicable because anxiety is fear in the face of *possibility*—both the threatening and the attracting aspects of possibility (rather than fear of a perceived danger)—and that therefore anxiety involves imagination, and is a purely human rather than an animal emotion, Heidegger thoroughly confuses the issue by saying that fear is "merely empirical" whereas anxiety is "ontological," (meaning "more primordial,") and it even means that anxiety is "what first makes fear possible." (4) In this essay we have always emphasized that grownup human emotions are much more complex than the emotions of children and animals, complicated by the coordinate development of such modifying factors as imagination, memory, anticipation, self-consciousness, world consciousness, personal interactions with others, and the much wider scope of the world in which grownups have to live and act. Thus there is no excuse for any reductionist view of human emotions. But it does seem a bit like *putting the cart before the horse* to say that the grownup, complex human emotion, anxiety, is what makes the simple, animallike, childlike emotion, fear, *possible.* Yet we read "Only because anxiety is always latent in Being-in-the-world, can such Being-in-the-world, as Being which is alongside the 'world' and which is concernful in its state-of-mind, ever be afraid." (5) Lucky Cro-Magnon man, to be equipped with latent-anxiety-over-the-atom-bomb, otherwise he never could have felt the appropriate fear-of-the-bear!

An even more confusing aspect of Heidegger's "method" is his manner of choosing his descriptive words, which, we must assume, were not dictated to him. He seemed to have a secret horror of evaluation, as though that were not part of existence and part of every writer's task, and as though it were any recommendation for an existentialist writer to pretend to a quasi-scientific "objectivity." *Being and Time* is sprinkled throughout with pejorative and commendatory words, accompanied by flat denials that any "evaluation" is being done. Thus the entire presentation of Dasein's *everydayness* is in such terms as "idle talk," "gossip," "chatter," and the average, ordinary view of life that has achieved the "publicness" of the "they" understanding of things. Everydayness is also characterized by a floating, unattached, groundless curiosity, which is not interested in knowing or wondering, but only in the "distraction" of "novelty for novelty's sake" and in providing idle talk with its subject matter. Also by ambiguity, the very essence of "the way things have been publicly interpreted." Idle talk, curiosity, and ambiguity are interconnected:

> In the ambiguity of the way things have been publicly interpreted, talking about things ahead of the game and making surmises about them curiously, gets passed off as what is really happening, while taking action and carrying something through gets stamped as something merely subsequent and unimportant. Thus Dasein's understanding in the "they" is constantly *going wrong* in its projects as regards the genuine possibilities of Being. Dasein is always ambiguously "there"—that is to say, in that public disclosedness of Being-with-one-another where the loudest idle talk and the most ingenious curiosity keep "things moving," where, in an everyday manner, everything (and at bottom nothing) is happening. (6).

In these three elements—idle talk, curiosity, and ambiguity—there is revealed the basic kind of Being that belongs to everydayness; Heidegger calls it "falling" or *verfallenheit*. But no matter how much we may wish to *agree* with this evaluative description of everydayness, including the negative connotations of "falling," we are next informed that

> ["Falling,"] does not express any negative evaluation, but is used to signify that Dasein is proximally and for the most part *alongside* the "world" of its concern. This "absorption in . . ." has mostly the character of Being-lost in the publicness of the "they." Dasein has, in the first instance, fallen away from itself as an authentic potentiality for Being its Self, and has fallen into the "world." "Fallenness" into the "world" means an absorption in Being-with-one-another, in so far as the latter is guided by idle talk, curiosity and ambiguity. Through the interpretation of falling, what we have called the "inauthenticity" of Dasein may now be defined more precisely. On no account, however, do the terms "inauthentic" and "non-authentic" signify "really not," as if in this mode of Being, Dasein were altogether to lose its Being. "Inauthenticity" does not mean anything like Being-no-longer-in-the-world, but amounts rather to a quite distinctive kind of Being-in-the-world—the kind which is completely fascinated by the "world" and by the Dasein-with of Others in the "they." Not-Being-its-self functions as a *positive* possibility of that entity which, in its essential concern, is absorbed in a world. This kind of *not-Being* has to be conceived as that kind of being which is closest to Dasein and in which Dasein maintains itself for the most part. (7)

All this is reminiscent of Kierkegaard's treatment of the "crowd" as "untruth" and of the sphere of "immediacy" in which most of mankind spends its days. It will be recollected, however, that in *The Sickness unto Death* Kierkegaard treats not-willing-to-be-oneself as one of the forms of despair, the sickness unto death, and makes no bones about *that* being an evaluative designation. (See my *In Search of the Self: The Individual in the Thought of Kierkegaard*, Chapter X. See also, for a comparison with Heidegger's treatment of fear and anxiety, Chapter IX, Anxiety, the Exile from Eden.)

As for the negative connotations of fallenness, Heidegger hastens to add, almost as if to forestall any Christian or Hebrew interpretations: "So neither must we take the fallenness of Dasein as a 'fall' from a purer and higher 'primal status.' Not only do we lack any experience of this ontically, but ontologically we lack any possibilities or clues for interpreting it." (8) And even the psychotherapist's well-meant efforts to help the patient attain authenticity in his existence are, as it were, rejected in advance: "We would also misunderstand the ontological–existential structure of falling if we were to ascribe to it the sense of a bad or deplorable ontical property of which, perhaps, more advanced stages of human culture might be able to rid themselves." (9) One would think that if Heidegger was able to foresee such misunderstandings of his evaluative-sounding nonevaluations, he would have been able to find, somewhere in the language, some more neutral-sounding terms than "fallenness" and "inauthentic."

From the standpoint of criticizing his effort as a philosophical anthropology, we cannot escape the evidence that Heidegger simply did not face the *fact* of evaluation, the part played by evaluation in human existence, whether authentic or inauthentic, and the relationship between evaluation, possibilities, and action. He instead provided himself with a convenient "rug" under which to sweep anything that he did not wish to face or to take seriously—namely, "idle talk and the way things have been publicly interpreted." If there is something "inauthentic" about evaluation, why not tell us what it is, and how it is got around and escaped from in "authentic" existence?

Heidegger continually talks about the *genuine* potentialities of Being. But let us dare to approach more closely and to ask persistently, what *are* such genuine possibilities? Could you not please give us even one example, one instance of a genuine possibility (other than death) for a particular Dasein? We search in vain through the length of *Being and Time* for an answer. (Compare the situation of the therapist, who certainly must face this question with each particular patient.) It is now as if, by the method of "making transparent," by "seeing through" the "merely empirical" to the "primordially ontological," we were beginning to catch sight of the emptiness, the vacuity, and the nullity of the concept of pure being, and, not surprisingly, also of the potentialities of this empty concept. For it is the concept of pure being that Heidegger is after in his fundamental ontology, using Dasein as a means to that end, not the idea of being as the fullest concept in the much more fuzzy but concretely overwhelming notion of what-is-in-totality. *That* would lead to metaphysics, which he has declared impossible on the merely empirical, cognitive level, whereas we have said that it is, on that level and with all due diffidence and precautions, at least approachable. In the essay "What is Metaphysics?", Heidegger insists that

> As certainly as we shall never comprehend absolutely the totality of what-is, it is equally certain that we find ourselves placed in the midst of what-is and that this is somehow revealed in totality. Ultimately there is an essential difference between comprehending the totality of what-is and finding ourselves in the midst of what-is-in-totality. The former is absolutely impossible. The latter is going on in existence all the time. (10)

Here it is as if he were hovering on the very brink of acknowledging the error of absolutistic claims for human knowledge, and of discovering the Ultimate Mystery. But no such illumination greets us in *Being and Time*, where the author tends rather to undulate ambiguously between *being* used as an equivalence category (the most vacuous category because its only criterial attribute is "is") and *being* used as an identity category (the *name* for what-is-in-totality, or the fullest category). If he had used the *latter* to characterize Dasein, he would have had to call it something like "premonition-of-totality-in-the-world," but he did not, for he had not stopped thinking that pure being is a "something" that belongs to entities.

In *Being and Time,* the emptiness, vacuity, and nullity of Being do not really become visible to us, in Heidegger's "transparent making," until we come to his ontological interpretation of conscience. It will be observed by the careful reader that as he moves from the front to the back of *Being and Time*, while Dasein moves from fallenness in "everydayness" to his authentic self in "Being-toward-his-ownmost-potentiality," Dasein inhabits an increasingly shrinking, isolated, and even solipsistic world. We cannot help wondering what good it would do, for example, for the poets of the future to "name what is holy." Surely in his Eigenwelt *authentic* Dasein hears no poets, past or future, nor does he hear any pronouncements of ethics, religion, politics, or philosophy—all such utterances belong either to the Mitwelt, or to the inauthentic, "public" world in which the "they" self is lost. And the very first "call" of conscience is: "Conscience summons Dasein's Self from its lostness in the 'they.'" (11) Even language, privately used, as in the interior monologue, is too public an instrument; hence "*Conscience discourses solely and constantly in the mode of keeping silent.*" (12; italics his) The mood of anxiety, especially in the face of "*my* death," individualizes Dasein to the point of feeling its uncanniness as "thrown possibility."

In the face of its thrownness Dasein flees to the relief which comes with the supposed freedom of the they-self. This fleeing has been described as a fleeing in the face of the uncanniness which is basically determinative for individualized Being-in-the-world. Uncanniness reveals itself authentically in the basic state of mind of anxiety; and, as the most elemental way in which thrown Dasein is

> disclosed, it puts Dasein's Being-in-the-world face to face with the "nothing" of the world; in the face of this "nothing", Dasein is anxious with anxiety about its ownmost potentiality-for-Being. (13)

And again, "In its 'who', the caller is definable in a 'worldly' way by *nothing* at all. The caller is Dasein in its uncanniness: primordial, thrown Being-in-the-world as the 'not-at-home'—the bare 'that-it-is' in the nothing of the world." (14) Ah, but if Dasein's basic unity is "Being-in-the-world," what is there left of the Being when the world is revealed as *nothing*? You guessed it—Being is revealed as nullity and the basis of a nullity, whose unity we might paraphrase as Nullity-in-the-Nothing. And *that* is the ontological definition of *guilt*: being the basis of a nullity.

> In the structure of thrownness, as in that of projection, there lies essentially a nullity. This nullity is the basis for the possibility of *in*authentic Dasein in its falling; and as falling, every inauthentic Dasein factically is. *Care itself, in its very essence is permeated with nullity through and through.* Thus "Care"—Dasein's Being—means, as thrown projection, Being-the-basis of a nullity (and this Being-the-basis is itself null). This means that *Dasein as such is guilty*, if our formally existential definition of "guilt" as "being-the-basis-of-a-nullity" is indeed correct. (15)

Whoever understands the call of conscience correctly, understands it as "*wanting to have a conscience*" (16) and choosing "having-a-conscience" as "Being-free for one's ownmost Being-guilty." (17)

And now Heidegger seems almost proud that his ontological "seeing through" has produced a definition of conscience that is not only not in harmony with, but in direct conflict with, the "ordinary view," because the latter "sticks to what *'they'* know as conscience, and how 'they' follow it or fail to follow it." (18) And it does not bother him one bit that in so isolating Dasein from the outside world, from all possibility of communication with other Daseins, from God, and even from any God-surrogates, he, Heidegger, has reproduced the situation of insanity and thus put Dasein's wanting-to-have-a-conscience on a par with Dasein's wanting-to-be-Napoleon. Heidegger's primordial anthropology should not be criticized for being secular or pagan—there is nothing wrong or remarkable about that—but for driving precisely *authentic* Dasein to the brink of insanity.

Now this distressing outcome of Heidegger's anthropology (distressing, that is, for those psychotherapists who, disenchanted with repressed-rat views of man, snatched at it as a means of bringing hope for the future into the lives of their patients) was greeted with a most peculiar gleefulness in the ranks of certain Barthian theologians, for *they* saw in the nullity and nothingness that is at the basis of Dasein the reason for the "turning point" in Heidegger's own thought toward a different kind of thinking altogether—namely, "primal thinking" that would "overcome" metaphysics by getting "behind" it or "prior" to it. We cannot get involved in a theological argument here, for this is an essay on epistemology, but we must briefly examine the "turning point" to see if it indeed represents any change in Heidegger's epistemology, and eventually, anthropology.

The "nothing" at the basis of Dasein, claimed the Barthians, is the inevitable outcome of interpreting Dasein in terms of "nothing but" its own structure. "Put otherwise: To understand *Dasein* as 'thrown' does not relate it to a 'thrower' outside *Dasein*, but rather relates it to Dasein's own projection of itself. Dasein is grounded in nothing outside itself." (19) This very "nothing" is stated by Heidegger in "What is Metaphysics?" to be the proper subject matter of metaphysics, in contrast to the what-is which is the proper subject matter of the several sciences. For if the "meta" in metaphysics means "beyond all beings" (in the conceptual schema being/nothing) what else can be beyond all beings except that which we call nothing?

Dasein, held out into nothing, is beyond all beings, and has in this sense attained ultimate transcendence, the goal of metaphysics. Transcendence beyond all things leads neither to God, nor to the cosmos as the sum total of all beings, nor to an established Cartesian subject upon which a world of objects can be built, but rather to nothing. Thus the metaphysical question is answered in "nothing", and this anwer to the metaphysical question is at the same time the end of metaphysics." (20)

It is not, however, the end of thinking about the Being/nothing alternative for Heidegger; and the "turning point" does not come until after the "end" of metaphysics, when what amounts to a philosophical "conversion" occurs:

> The nothing that emerged when metaphysics sought to ground *Dasein* in something outside itself ceases to emerge as nothing, and instead being dawns. If it was metaphysics' engrossment with analyzing beings that prevented it from catching sight of Being, then the arrival at nothing, by ending the engrossment with beings, corresponds to the unveiling of being. (21)

It certainly does seem that somewhere between *Being and Time* and the essay "What is Metaphysics?" Heidegger had either read Rudolf Otto's *Idea of the Holy* or had experienced personally the awe aroused by the *mysterium tremendum et fascinans*. Then, apparently, he began to look around for a means of escape from the conceptual being/nothing strait-jacket. But old habits die hard, most of all intellectual ones. Thus even when, at the end of the essay, he proposes a "leap" reminiscent of Kierkegaard's "leap into the arms of God" (a matter of faith, not knowledge) he is still in the same conceptual groove:

> Philosophy is only set in motion by leaping with all its being, as only it can, into the ground-possibilities of Being as a whole. For this leap the following things are of crucial importance: firstly, leaving room for what-is-in-totality; secondly, letting oneself go into

Nothing, that is to say, freeing oneself from the idols we all have and to which we are wont to go cringing; lastly, letting this "suspense" range where it will, so that it may continually swing back again to the ground question of metaphysics, which is wrested from Nothing itself:

Why is there Being at all—why not far rather Nothing? (22)

Does this mean that Heidegger is moving around toward a more nearly biblical view of man? Yes, in the sense that the motion is toward *Homo religiosus*, special variety, intellectual mystical contemplation. He seems to be glimpsing the Ultimate Mystery, but not yet its transcendence of human thought schemas, including the Being/Nothing schema, or any "primal thinking" schemas yet to come. Naturally he turns for help to the emotions, or "moods," as he calls them, for help. Where else?

> This turn in direction brings with it a reversal of the basic mood of Heidegger's philosophy. Rather than calling man 'the one who stands in nothing's place,' Heidegger now speaks of him as the 'shepherd of being.' Instead of anxiety, there emerged gratitude for being's 'favor.' Once the Promethean direction of metaphysics is renounced, the positive emerges (23).

Here we must leave the later Heidegger, still engaged in "primal thinking," and having a long, long road ahead of him, as all will know who have explored along the many paths of contemplative mysticism. *That* there is a Mystery we would be the last to deny, but that it can be comprehended by man in any human terms whatever simply means, in our epistemology, that it has not yet been reached. (It is, in terms of our crude map, *beyond* perception, conception, emotion, and imagination.) Yet Heidegger said, "The thinker utters Being. The poet names what is holy." (24) Indeed, thinkers and poets since the dawn of human consciousness have stood before the Mystery, and within the Mystery, completely surrounded and interpenetrated by it, and they have uttered and they have named; but their speech on those occasions has been an act of faith, not of knowledge. What choice has man, confronted by the Mystery, but to select from what is knowable that which seems to him the most important, and to extrapolate it, in *faith*, into the Unknowable? And that is precisely where the totality mapping comes in, not only for the different religions and philosophies but also for individuals as naive or sophisticated ontologists, to influence their choosing of what is most important within their most inclusive view of reality.

But where does that leave the psychotherapist who wants to use Heidegger's *Daseinanalyse* to bring hope for the future into his patients' lives? I am afraid it leaves him where he has always been—on that uncomfortable spot of having either to give the patient his own philosophy—what it is that keeps *him* going—or else to show the patient how he can work out a philosophy of his own. First the therapist must examine himself, determining whether *he* really lives according to the heroic nihilism of self-assertion in realizing his possibilities over against "nothing," and if not—must be then not turn in the same direction as Heidegger, and perhaps even recommend *that* to his patients? But psychotherapists are busy, practical men, not philosophers, and so are their patients. And these are such lean, hungry times among the official humanistic philosophers, teetering (from starvation?) between scientism and nihilism. So perhaps I may be forgiven for recommending the biblical view of man as delineated in this essay as a third alternative, or at least as a "corrective" for some of the more obvious omissions in what we have been told is the *structure* of Dasein itself.

In the terms of Heidegger's own hyphenating game, the very least correction that would be necessary would be to change Dasein's basic unity from Being-in-the-world to *Becoming*-in-the-world, to stress the temporal, biographical, developmental stretching-out, and the co-presence of past-present-future in every moment of Dasein's *there*. Becoming does not so easily melt into nothing as Being. But even that would be too static, too much a structure, too boxlike, to do justice to the human type of becoming, which is no process-in-a-box. For it does not show at all the reciprocal interaction between each Dasein's becoming and his world, the receiving and giving, the message-taking and responding, the venturing-forth and discovering, which modify both the Becoming and the World. So we should have to call Dasein at least Becoming-through-reciprocal-interaction-with-the-world. But even that is not enough, for it does not show the growth on both sides (and the possible arrest), nor the direction of the growth. Well, then, how about at least Becoming-toward-totality-through-imaginative-reciprocal-interaction-with-an-expanding-world? No need to worry about reaching totality, for it is only a direction, nor about losing individuality, where so many choices are involved.

But I think I have made my point. The complexity of human becoming will never be captured in a hyphenated phrase; and what is more important, the patient sitting in front of the therapist is not a hyphenated phrase, but a breathing, suffering human being. The only excuse for conceptual schemas or "models" is that they should be helpful to man, the animal that is most in need of help, caught between inhuman evolution and far-from-human society.

To those psychotherapists who find the Heideggerian *Daseinanalyse* helpful, I should like to suggest two further lines of action. The first is to reexamine his entire treatment of the emotions, with a view to answering such questions as: Is it actually true that *only* in anxiety, and *only* in anticipation of my death do I become aware of my own individuality? Is it not rather the case that in my entire emotional life as extended through imagination, and in my interactions with others, yes, even in those casual interactions of inauthentic "everydayness," I become aware of my differences from others, as well as similarities, and also of *their* differences from me—*their* individualities? Do I not feel, the more I exercise empathic imagining, that every person is unique, and do I not know the pathos of time-running-out for each one? Here I perceive in Heidegger a deficient understanding of empathy. In fact he understood empathy as a "first bridge," ontologically, from one Dasein to another, but only in the most minimal sense that one understands that the Other is not a *thing*, but another *like* oneself. "The relationship-of-Being which one has towards Others would then become a Projection of one's own Being-towards-oneself 'into someone else.' The other would be a duplicate of the Self." (25) Now really, if that were true, how could the parent understand the child or the teacher the pupil? How could the psychiatrist empathize the world of the psychotic? And how could the novelist create a novel with thirty-five different characters, not one of them a "duplicate" of his own Self? As for the analysis of anxiety that results from "seeing through" the merely empirical "fear" to the ontologically more primordial "threat of nothingness," while I freely confess that I do not possess this ontological x-ray vision, I cannot restrain a certain curiosity about what would happen if someone so gifted were to train such vision on the other seven members of the most primary emotion spectrum, at nearly the same intensity level as fear. What would be the more primordial, ontological versions of sorrow, disgust, anticipation, anger, joy, acceptance, and surprise? Perhaps there might be some hope there for poor Authentic Dasein, to keep him from going crazy in his existential solitary confinement with Nothing.

Second, I would suggest a much more concrete spelling-out of what it is that Heidegger really means by *potentia*, potentiality, possibility, and so on. There are about fifty-five phrases containing these words in the index of English expressions at the back of *Being and Time*; yet the concept remains strangely abstract and devoid of specific content each time such a phrase is used. Here we must remind ourselves that, whereas in the physical sciences abstraction is a virtue when firmly connected to a "model," there is a tendency, in anything dealing with the specifically human, for abstraction to be spuriously universal, vacuous, and hence delusive; and if overtolerated as harmless or "objective," it might even lead to insanity. In conceiving possibilities, imagination must be used; hence opportunities for vain imaginations abound, the more so, the more vague or emotionally inflated the concept is allowed to remain.

We must try to pin it down with questions. Heidegger says that death is the possibility of the end of all possibility, the one "genuine" possibility whose only uncertainty is the time and place of its occurrence. Does he mean that one's ownmost possibility ends with one's own death? If I invent the atom bomb, or a cure for cancer, and then die peacefully in my bed, does my potentiality end right there? In that case, why is history cluttered with the still unfolding harmful and helpful possibilities initiated by people who died hundreds of years ago? Or, suppose I just "sit on my hands" and do nothing until I reach the age of 90. Does that then count as my ownmost potentiality—living for 90 years? Like Aristotle's acorn becoming an old oak tree? Is Heidegger's Dasein every existing individual, without exception, or only some strangely exceptional or unusually talented person? Is Dasein male or female? How old is Dasein? And what does Heidegger mean by "fetching back" what was once a genuine possibility for someone in the past and repeating in the present? How do I know it was "genuine" for him, and even if I assume this, how could it be ownmost for me if it had been ownmost for him? Is history repeatable? And what about all the possibilities I have to refuse for every one that I actualize? And what about the possibilities that I see, too late, that I should have refused? And why all this agonizing in the face of Nothing—why not just flip a coin?

The later Heidegger no longer talks about Aristotle taking a "wrong turning" in the direction of things as against persons. He talks rather about the whole history of Western philosophy from Plato to Nietszche and modern technology having been "led astray" by "being" in a "fate-laden" movement:

> "There is being only from occasion to occasion in this and that fate-laden contour: *physics, logos, hen, idea, energeia*, substantiality, objectivity, subjectivity, will, will to power, will to will." This series of interpretations of being reflects the occurrence of being from early Greek thought via Plato and Aristotle to the Middle Ages, Descartes, German idealism, Nietszche, and finally modern technology. (26)

Far be it from me to underestimate the difficulties of thinking about Being! I can only leave this summary as a suggestive warning to those who are interested in rescuing *Daseinanalyse* from a certain "fate-laden" vacuity inherent in the abstractness of its concepts, especially the concept of *potentia*. I can only plead, wherever persons are concerned, as against things, concretize the concept! Give examples! Tell stories! Recite parables! Reenact histories!

REFERENCES

1. R. G. Collingwood, *An Essay on Metaphysics*, Oxford University Press, London, 1940, p. 14.
2. Martin Heidegger, *Being and Time*, Transl. John Macquarrie and Edward Robinson, Harper & Row, Publishers, Inc., New York, 1962, p. 27.
3. Heidegger, *Being and Time*, p. 35.
4. Heidegger, *Being and Time*, p. 230.
5. Heidegger, *Being and Time*, p. 234.
6. Heidegger, *Being and Time*, pp. 218–219.
7. Heidegger, *Being and Time*, p. 220.
8. Heidegger, *Being and Time*, p. 220.

9. Heidegger, *Being and Time*, p. 220.
10. Martin Heidegger, "What is Metaphysics?," in *Existence and Being*, Introduction by Werner Brook, Henry Regnery Company, Chicago, 1949, p. 363.
11. Heidegger, *Being and Time*, p. 319.
12. Heidegger, *Being and Time*, p. 318.
13. Heidegger, *Being and Time*, p. 321.
14. Heidegger, *Being and Time*, p. 321.
15. Heidegger, *Being and Time*, p. 331.
16. Heidegger, *Being and Time*, p. 334.
17. Heidegger, *Being and Time*, p. 334.
18. Heidegger, *Being and Time*, p. 335.
19. James M. Robinson and John B. Cobb, Jr., Eds., *The Later Heidegger and Theology,* Vol. I, *New Frontiers in Theology,* Harper Row, Publishers, Inc., New York, 1963, p. 11.
20. Robinson and Cobb, *The Later Heidegger,* pp. 11–12.
21. Robinson and Cobb, *The Later Heidegger*, p. 12.
22. Heidegger, "What is Metaphysics?," in *Existence and Being,* p. 380.
23. Robinson and Cobb, *The Later Heidegger*, p. 13.
24. Heidegger, "What is Metaphysics?," in *Existence and Being*, p. 391.
25. Heidegger, *Being and Time*, p. 162.
26. Robinson and Cobb, *The Later Heidegger*, p. 27.

Twelve Religious and Theological Implications

The crude map of the human type of consciousness (Figure 1*a* on p. 151) yields a general philosophy of religion, since it provides the raw materials that must be there and shows what man is doing with them, when he is behaving religiously. The permanent background of impenetrable mystery is the utterly transcendent aspect of God. This is the goal of all true mysticism—to approach the Mystery as closely as one can, to contemplate it, to become united with it. Human cognition is precisely what it transcends, as the mystics of the *via negativa* keep telling us. It is not this, not that, and if you can even say "boo" about it, you have not yet reached it. It is the Divine Darkness of St. Denis, and the *coincidentio oppositorum* of Nicolas of Cusa. All opposites are swallowed up in it, not (as Hegel mistakenly thought) because they are synthesized, but because all opposites are equally applicable and equally inapplicable. In Eastern mysticism, it is the awareness of the All in which the human soul as a separate thing disappears like a drop of water in the ocean. Yet some degree of awareness as a noncognitive feeling remains—the "oceanic feeling." This in turn gives rise to the *via positiva*, to saying "thou art that" to everything, including the divine within the soul, since this makes as much sense as "thou art not that," which in turn leads to accusations of self-deification against mystics in their moments of ecstatic union. For instance, the Quakers when they are quaking. Or Jesus when he said, "I and the Father are One." Blasphemy! Nirvana is also described as a state of bliss, but it is more like a feeling of emancipation and release from the sorrows of existence as an individual and a body. Obviously there is a wide borderline area between the transcendent and the immanent, depending on the credence given to dreams, visions, voices, emotional exaltations, attention-paying and witholding, and expectation of results, in different cultures. But the experience of the presence of the Holy is the grass roots, the raw materials, of all religions; it is that to which they must return when they become unsure or confused in their more specific philosophical or mythological teachings.

Insofar, then, as anything is said about God in words, it is said in the realm of immanence, even if it "points to" a God that is understood to be transcendent. In the context of the Judeo-Christian tradition, the "image of God in man" would be man's awareness of totality, no matter how dim or inchoate. "Original sin" would be the failure to transcend the natural childhood selfishness, necessary for survival, in the direction of a more inclusive reality—first family, then tribe, then humanity, ultimately totality, including past, present, and future. A Quaker has put it well: God is the Beyond that is within you. But it is precisely the transcendence of the beyondness within that raises the question of what to do with the immanence that is also within and all around us, and not for Christians alone. This leads to the quest for revelation, for a teacher, prophet, Bodhisattva, charismatic person, or saint, who seems to be in touch with the beyond and yet can advise on how to behave in the here-now.

So-called primitive mythologies are the cosmologies of a particular tribe or people. They both describe and explain that particular tribe's totality-view of the universe. They are intentionally ethnocentric, beginning with origin stories for the whole world and the special favor of some god or goddess in giving this people its land or hunting rights or special privileges. Every primitive tribe is the chosen people of some god or goddess, and the more closely we approach Stone Age cultures, the more the tendency is for the tribe to regard itself as generic man and all other human forms as gods or devils. In addition to origin stories, there are the important events in the history of the tribe, which are memorialized by being turned into legends and hero stories. "So it is in the legends of my people" is the beginning of history in every tribe, for without being turned into legend, the events themselves would have been neither remembered nor retold.

The gods and goddesses of a given mythology are the personified powers of nature and history that are regarded as real, more powerful than man, and as having man at their mercy. Whether friendly or unfriendly, they must be worshipped, for that is the only chance of gaining their favor or escaping their wrath. Not only do the gods satisfy man's curiosity about the world's constitution, they provide an ever-ready explanation for the success or failure of particular human enterprises. The gods are on the side of power and success, and it is for power and success that they are petitioned, invoked, praised, and given thanks. A defeated tribe will often adopt the gods of its conquerors, since its own have just proven themselves powerless. Unfriendly gods must be honored even more than the friendly ones—if they are not mollified, venerated, and propitiated with offerings, one courts disaster.

The change from polytheism to monotheism represents man's increasing consciousness that it is *of the totality* that polytheism is attempting an explanation. Monotheism raises more problems than it solves, since to the single God must now be attributed all the conflicting powers of nature and history (some friendly and some inimical to man) that formerly were so easily parceled out among conflicting gods and goddesses. Solving the problems of monotheism is the hope and despair, the lure and the stumbling block, of metaphysical and theological thinking. There can be only one totality, but within it there must be found a way to represent and explain the diversity, the conflict, the good and evil, and the hope of deliverance or escape that characterize man's actual life in the world. In general, polytheism is left to the masses as a "consolation" while a small spiritual élite tries to work out the doctrinal and ethical implications of a monotheistic leader or leaders. This is how we arrive at the actual "higher" religions of mankind, some more philosophical, some more mythological, in their expression and explanation of man's ultimate concern, totality.

From the theory of knowledge set forth in this essay, it is hardly to be expected that there will ever be a monotheistic system of thinking that is both logically and psychologically satisfactory. (I am counting the Hindu Brahman–Atman relationship as a philosophical equivalent of the monotheistic God–soul relationship.) Because of selective distortion and systematic distortion involved in language itself, because of the nature of the "referent"—totality—and also because of the "stage we are in" historically, monotheism will always be a matter of backing away from this pitfall to avoid some other pitfall only to fall into a third pitfall, and so on. Monotheisms are too fatalistic or too voluntaristic, too pantheistic or too other-worldly, too aggressive or too passive, too personal or too impersonal, too human or too inhuman, too near at hand or too remote, too eternal and unchanging or too evanescent and fleeting in the exalted "moment." Man cannot simply deify and worship his own awareness of totality—it is too huge, too diffuse, too multiple, too incomprehensible, and it contains too much that is not worthy of being worshipped. The function of the monotheistic leader is to select some qualities of the totality that are worthy of admiration, worship, and emulation, such as goodness, justice, power, deliverance, or love, and then to perform a theodicy that will persuade his followers that eventually these "positive" values are bound to win over the undeniable "disvalues," which must be explained away, battled, overcome, accepted, or just endured, for the sake of the selected positive values. But this very process of *selection* constantly threatens monotheism with breakdown, into at least ditheism, such as we see in the Ahuramazda–Ahriman of Zoroastrianism, or the popular mythology of the "good guys" versus the "bad guys." Thus the basic problem of monotheism is theodicy: accounting for the disvalues after the highest values have been selected. Polytheism knows no problem of theodicy; it merely threatens chaos.

The modern scientific–technological world view is, by a curious coincidence, much closer to the polytheistic mythologies of primitive peoples than to any of the monotheistic higher religions. This is because, although science claims to use the same *method* everywhere, the subject matter to which it is applied offers different degrees of recalcitrance to the method and imposes different kinds of model making and omission to get any predictable results at all, thus issuing in the multiplicity of technologies we have today. The analogy between "white man's magic" and so-called primitive magic is too true to be good: in return for certain ritual offerings and performances, the petitioner is promised that his desires will be fulfilled, such as long life or the death of an enemy. Technology attempts no totality view. It dispenses power to those who are willing to pay the price.

But in the scientific theories of cosmology plus the biological theory of evolution we have modern scientific attempts to describe and explain the totality of the universe; therefore these efforts are true *competitors* to the religious and metaphysical cosmologies of the past. Very few of the religious cosmologies of the higher religions have tried to come to terms with scientific cosmology and biological evolution, because the sheer enormity of the time scale and space scale of the modern universe makes the religious world views seem ridiculously small and provincial. Mostly they have side-stepped the issue by saying that science deals only with matter, whereas religion deals with the spiritual nature and destiny of man. We could easily dispose of the pronouncements of astronomy or physics in this way, but biological evolution presents a thornier obstacle, especially in the Judeo-Christian tradition, which gave rise to evolutionary science. No matter how tolerantly and broadmindedly we now demythologize and deliteralize the creation stories in Genesis, it is certainly hard to square off the idea of a loving Father-creator with anything as utterly inhumane and indifferent to suffering as the process of evolution. Neither Christian nor Jewish theologians have really faced up to the moral implications of evolution as a *method* of creation. Animals (and small children) can be exonerated because they are unconscious of the suffering they inflict in performing their survival patterns; but no such excuses can be extended to grownup men and women, who know very well when and where they are causing suffering to others and in some cases, positively enjoy the feeling of *power* derived from the ability to make others suffer.

Furthermore, no matter what heights of spiritual emancipation the parents as individuals might attain, their children, thanks to the noninheritance of cultural characteristics, have to start all over again as Stone Age babies: somehow they must recapitulate the whole human-making process by passing rapidly through a series of behavior-molding institutions. The usual double aspect of threat and promise is presented by this return to the culturally "clean" genetic base with each new generation of children. It is a chance for a new beginning and a correction of cultural errors of the past; but it is also a chance for losing the cultural gains of the past. This closeness to the genetic biological basis of behavior, and the constant reversion to it with every new generation, gives a "thin veneer" aspect to all cultures. To those highly developed cultures that cannot be transmitted without many years of education, it gives a luxury aspect—only the members of an élite class or of a wealthy society can afford it. People living on the very margins of survival have no choice but to revert for mutual help and protection to the family and tribal arrangements that served for survival in prehistoric times and even among the primates. To appreciate how "thin" the cultural veneer is, temporally, we must remember that we are all only about 15 acculturation years away from the biological beginning (15 being a generous estimate for the age at which present-day Stone Agers and Early Bronze Agers are considered grownups), and of those 15 years the first five are spent in a state of incapacity to survive without grownups. Our nearest primate cousins are grown up at 5 and ready to start new families at 8.

A great deal depends, both psychologically and theologically, on where we draw the line historically (between the human and the subhuman) and in contemporary life (between the child and the grownup). The people who wrote the creation mythology of the Old Testament no doubt regarded their polytheistic contemporaries as something less than human, not merely children, but "people who sit in great darkness." And indeed we could argue that biological evolution is the best theodicy of all, since it shows how much must have been lived through by the brain-evolving process before an animal capable of totality awareness became possible, let alone a monotheistic animal. Arend Th. van Leeuwen, in *Christianity in World History*, (1), argues that the second verse of Genesis, "And the earth was without form and void," should be understood as refering to the *chaos* of polytheistic primeval prehistory, from which God rescued Israel, rather than only the raw materials of the geophysical earth and sky. Such a view should at least make us more tolerant of some degree of chaos at all times, since we are always surrounded by various hangovers and by-products of the evolutionary process.

All people are in different stages of the human-making process, and many of them are incapable, either genetically or culturally, of going very far in that direction, even in a lifetime. They remain as (grownup) adolescents all their lives, responding with the proper childhood ethic of "seek pleasure and avoid unpleasure" to whatever their society offers to grownups. In a wealthy, self-styled "free-enterprise" society such as the American (which to non-Americans always looks more like the freedom to be as greedy as the law will allow) whole economic institutions pander to this rock-bottom motivation in the supposed grownups with money to spend—spurred and prodded by the systematic lying-for-profit of the Madison Avenue hucksters. When and where can we raise the question of how good for the grownup, or for the process of growing up, is the constant maximization of pleasure and minimization of pain and effort, especially insofar as this policy demonstrably leads to satiation and boredom rather than bliss? Tired of candy and toys? Try these new brands of candy and toys. Instantly the economic system produces whole industries to service those who are satiated and bored, thereby producing even more. . . . We are like fat, glutted children on an assembly line for producing fatter, more glutted children, and we can't get off because this assembly line keeps the wheels of industry humming and the unemployed off the streets, proving to the world what free enterprise can do. But what about the freedom to say pardon me, enough is enough, I happen to have some growing up to do? And what about the communist countries to whom we are proving something? The same thing will happen to them as soon as they catch up with us on consumer goods, especially candy and toys. Their temporary asceticism and idealism will soon evaporate, as soon as satiation and boredom begin to show. It is one thing to sacrifice for your brother who is starving, and something else to lay down your life for better refrigerators in Novosibirsk. Let them use ice. What a relief it would be to get out of this treadmill. There must be something wrong with the hedonistic ethic for grownups.

In our study of the animal value system, we noted that it operates preservatively for the animals and for children precisely because it deals with immediate situations and short-range goals. No "ordering of priorities" is needed by any animal because the first priority—survival to the point of reproduction—is served by each primary emotion in its proper animal-existential situation. In contrast, not only must the human child survive through a much longer period of helplessness and dependency on grownups, he is subjected to an acculturation process that imposes on him the value priorities of his particular tribe or society. But why value priorities at all? Why does the animal value system *not* work for grownups, just as it is? It is because of the switch from short-range goals to long-range goals that is brought on by the time-stretching of human consciousness to include the past by imaginative recollection and the future by projective imagining. No animal has to plan his day's activities, or have them planned for him by a grownup, as does even a preschool child. As soon as a plan is involved, there is a series of goals arranged in a means-to-end logic. The final end may be made definite or left vague. But even a vague goal must be sufficiently attractive to exercise a constant pull on the present. Feedback is required—a constant reinterpretation of the past, with respect to how well it did or did not support one's efforts to reach intervening goals. Whether for freely chosen ends or socially imposed goals, whether through private reflection or cultural "conditioning," and whether through hope of imagined reward or fear of imagined punishment, somewhere between the ages of 3 and 10 years, the human child is turned into a teleological animal. Even if the hedonistic ethic is retained, it becomes prudential: how to get from A to K through the intervening points $B, \ldots, J$ with the least amount of pain or punishment and the greatest amount of pleasure.

Here, in the teleological structure of human existence brought on by long-range planning, lies the origin of man's "freedom," as compared with animal existence. Far from being "free," as we enviously think of them, the animals are unconsciously "obeying" (i.e., responding to) the action-guidance pattern that has evolved by natural selection from the animal value system in their species. Yet even a 5-year-old human child, in his game playing, must be able to make plans and thus repress immediate responses for the sake of the goal envisioned in the game. Think of the repression of immediate response, such as fear or anger or laughter, and the concentration of attention, and the control of every action in relation to the goal, that are required, say, by a "primitive" hunting party, stalking an animal into a trap.

The change from short-range goals to long-range goals certainly deprives human existence of the charming spontaneity of childhood; but it enormously increases the efficiency, orderliness, directedness, and will-power of grownup life. The long-range goal focuses attention, eliminates distraction, concentrates curiosity on the means-to-end strategy, assigns "meaning" to every action in terms of leading toward or away from the goal. It heightens expectation, perseverance, and the excitement of living; and it lowers the feelings of satiation, boredom, lassitude, languor, somnolence, and laziness. Compare the household cat: when not *at the moment* pushed or pulled by hunger, sex, curiosity, or danger, she curls up in the most comfortable chair and goes to sleep.

Thus man, the teleological animal, tries to explain the world in which he pursues his long-range goals: is it any wonder that all his early explanations are teleological? What else could he use as his "model of explanation" if not himself? Powers of nature and history, yes—these he is surely surrounded by. But why personified powers? Why friendly and inimical powers? Why powers that must be praised and thanked, mollified and propitiated, invoked and bargained with? These questions lead straight into the demythologizing controversy, the death-of-God movement, and the supposed secularism of modern life.

Ancient scriptures are indeed ancient, couched in languages that "correspond" to prescientific world views and populated by agents and entities which, if taken literally, would make utter nonsense out of presently well-established cause-and-effect relationships. Since there is no getting around this, why bother with such writings at all? Why not relegate them to literary archeology? The answer to *this* question has something to do with "the stage we are in," historically. After two world wars, the atom bomb, the cold war, and the final threats to the species itself looming into sight, man's honeymoon with science is over. In digging into the ancient scriptures, we are no longer looking for cause-and-effect relationships; rather, we seek more clues to man's understanding of his own nature and destiny in history.

At first it was thought that the proper way to demythologize was to translate the "message" of the ancient scriptures into some idealistic philosophy, such as neo-Hegelianism, or Heidegger's existential ontology, or Whitehead's Process Philosophy. Unfortunately for this proposal, it turns out that even today the Ordinary Languages of the Bible, of Hindu Scriptures, of Buddhist teachings, of the Koran, or of Greek tragedy, make more sense to the "ordinary man"—and even to some scientists—than the abstruse juggling-with-vague-abstractions of idealist philosophers. We simply *must* ask: is it really true that modern, postscientific man *cannot* understand what the parable of the Good Samaritan, say, is trying to teach us, but he *could* understand it if it were all translated into neo-Hegelianism, neo-Kantianism, "primal thinking," or some form of Whitehead's metaphysics? We must leave that as a "way" for certain types of mind.

But fortunately there is another way. The object of demythologizing is to dispose of superstition and magic, while retaining whatever insights into the human condition a particular myth may contain. By using all the resources of art and literature, by imaginatively reconstructing the past historical situation and empathetically reliving it, we can see how certain beliefs were inevitable at certain times in history. For example, if we had no cure for a disease, and not the faintest notion of how to look for it, how else could it be fitted into our totality schema except as the work of enemy tribesmen, angry gods, or punishing fate? And how else could we respond except by calling for exorcism, repentance, sacrifice, or resignation? And are these actions and attitudes still not found today wherever the historical conditions are similar?

The specifically scientific objection to the mythology in the ancient scriptures is usually assumed to be that its language and concepts are too anthropomorphic. But we have seen in Chapter Seven that all forms of human knowledge *are* anthropomorphic, even where they are concerned with the outermost reaches of cosmology or the innermost fine structure of the atom. What are physicists doing when they talk about particles, waves, wavicles, waves of probability, or relativistic time-dilation, if not trying to construct, in terms of life on the human scale, conceptual models of things as they must be on the scales where the abstractions of mathematics lead us? And where is there a law (*there's* an anthropomorphism!) that says that structure on the very large and very small scale should bear any resemblance *at all* to structure on the medium scale (which man with his particular sensory-conceptual equipment occupies)? There is no law, only the faith of the insatiable-curiosity-driven scientist that, model making having taken us so far, why not farther?

What scientists should object to in the mythologies of the religious world views is not their anthropo*morphism* but their anthropo*centrism*—the fact that everything is explained teleologically in relation to the teleological center, man, in terms of his aims, his needs, his desires, his plans. It is a problem in excess, and of course it is the totality nature of the "referent" that leads to such excess. Strictly speaking, there should be no teleological explanations in regions of reality that were here long before man appeared on the scene and will still be here long after he is gone (when all other explanations will be gone also). Many scientists feel the disparity of this situation—the enormity of the universe, which nevertheless "exists" between the ears of a man and will cease to exist when the last human consciousness disappears. For this reason, many scientists are drawn to the Eastern type of mysticism, which stresses that man, a mere speck, is also in some sense the All. Erwin Schroedinger was referring to the Vedas when he wrote,

> This life of yours which you are living is not merely a piece of the entire existence, but is in a certain sense the *whole*; only this whole is not so constituted that it can be surveyed in one single glance. This, as we know, is what the Brahmins express in that sacred, mystic formula which is yet really so simple and so clear: *Tat tvam asi*, this is you. Or, again, in such words as "I am in the East and in the West, I am below and above, *I am this whole world*." (2)

But if the explanations of the world contained in mythologies were not anthropocentric at all, they would not be explanations for the explanation-seeking animal. Consider the three cases: lightning starting a forest fire; an arsonist starting a forest fire; and a fire-fighter starting a backfire to stop a forest fire. The first is now regarded as an impersonal power of nature, containing no hostile or friendly intentions. But are the second and third not more correctly regarded as a hostile power and a friendly power? To the degree that human intention enters, the powers that man is surrounded by are, by and large, friendly or hostile; therefore it is correct to personify them. *Magic and superstition* enter when we make wrong causal connections between what we can do to change these powers and what actually would change them. Casting spells to change the direction of the wind is superstition; starting a backfire to limit a forest fire is not. Rainmaking, the symbolic *reductio ad absurdum* of all religion by some scientists, is gradually moving from magic to technique.

Theologians (in their role as theologians, not as religious persons) must be to mythology as literary critics are to poetry;

and literary critics do not undermine or vitiate the purpose and goal of good poetry just because there are so many bad poems in the world! We know now that mythologies are works of art, just as stories, legends, and folklore are works of art. They may have started as the creations of individuals; insofar as they are repeated, elaborated, reenacted, and passed on from generation to generation by a tribe or a community, however, they are communal works of art. It should not surprise us—indeed, we should positively expect—that the kinds of feelings and life experiences that mythologies embody or enact or reenact are those feelings and life experiences that are *(a)* most important to and *(b)* most sharable by, the particular tribe or community that lives by them. Such feelings and life experiences tend to be the more generalized, publicly acknowledged feelings dealt with by the psychologist, but expressed in the garments of particularity of their communal origins. Just as magic is the science of the particular (Lévi-Strauss), so mythology might be called the psychology of the communal. Instead of denouncing every myth except our own as idolatrous, as if our own lives were not filled with unacknowledged idolatry, we should try to learn what myths tell us about peoples' loves, hates, fears, expectations, and hopes, both in so-called primitive and in so-called civilized societies. After reading this essay, is any one of my readers prepared to announce that the god Mars is dead? How about Eros and Aphrodite? Remember the population explosion! Well, what about Apollo? The god of reason, in our lifetime *he* made it to the moon—although here on earth he seems to be running wild, out of control, while Siva, the Hindu god of destruction, is crawling all over the planet, following in Apollo's footsteps. And Dionysius? As for redemption myths and savior myths, what do they tell us about what man is looking for?

In the context of Christianity and the death-of-God movement, we must first ask which particular concept or model of God was envisioned as the candidate for demise? Those which were nothing but magic and superstition have already died, having been replaced by white man's magic, technology. But even of those which were an attempted totality characterization, there are still plenty which deserve to die. For instance, the idea of God as an Omniscient, Omnipotent Monster, who visits natural disasters and undeserved suffering upon his children in order to improve their morals! (A woman in the latest Peruvian earthquake was seen to rush out of her house, throw her hands to the sky, and cry out: "Oh God, for what are we being punished?") It is not quite clear how the idea of God as omnipotent and omniscient gained ascendancy within the Christian tradition—perhaps it was the result of the Caesaro-papism that followed on the conversion of Constantine. But it is quite clear psychologically what a transparent piece of vain imagination and self-projection unto the divine it represents—*man* wants to be omnipotent and omniscient, and therefore the only kind of god he can worship has precisely the qualities of all-powerfulness and all-knowingness. But followers of Jesus believe that he is both the revelation and the incarnation of Divine Love. The only kind of God that Christians can worship is the God who *is* Divine Love. Is Divine Love omnipotent and omniscient? And does it send earthquakes to punish the wicked and torture the innocent? Much rethinking is needed here, and much jettisoning of antiquated ideas, of which the Church today is a veritable museum. If Christian theologians refuse to undertake this job, the cathedrals of Europe will become the Angkor Wat of Caesaro-papist Christianity—moldering stones, crawled over by tourists and archeologists. Meanwhile the followers of the living Christ may have to return to the modern equivalents of the pre-Constantinian catacombs and caves, living as small groups and congregations, prepared to be despised or barely tolerated while inaugurating the kingdom of Christlike love here and now.

Finally, who will demythologize the so-called secularity of modern life? Who will remove the superstition and magic by which it lives? If man is *Homo religiosus* for good or ill, is not secularity the greatest delusion of all? Are not all wars religious wars, since they assume that the top values of "my tribe" are worth killing and being killed for, compared with the top values of "your tribe"? Who will unmask the structure of tribalism in both its good and evil consequences, and the specifically Bronze Age tribalism armed with atomic spearheads that prevails in international affairs at the present time? And who will demythologize (remove the superstition and magic from) the tribalisms that characterize our innumerable subcultures, the organization of societies along the lines of class, age, race, education, religion, politics, professionalism, bureaucracy, sports, recreation, and so on, and so forth. Surely the animal that considers everything as grist for his religion-making mill—from science to politics to philosophy to art, from atheism to agnosticism to fanaticism—surely the animal who can say "more nihilist than thou" or "more despairing than thou" will not pass up the attractiveness in "more secular than thou"?

With this capsule "philosophy of religion," and with such questions, I must leave the reader, for the answers will depend not only on his own cultural commitments, but on the degree of conviction with which he has read the preceding essay on what is involved in knowing, doing, and surviving.

REFERENCES

1. Arend Th. Van Leeuwen, *Christianity in World History*, Charles Scribner's Sons, New York, 1964, pp. 64–66, passim.
2. Quoted in Jeremy Bernstein, *A Comprehensible World*, Random House, New York, ©1967, reprinted in *Project Physics Reader*, 5, "Models of the Atom," Holt, Rinehart & Winston, New York, 1968, p. 179.

Appendix One On Teilhard de Chardin's Cerebralization and Hominization

Since the present essay is so much oriented both along evolutionary and Judeo-Christian lines, it will probably be compared with at least some aspects of Teilhard de Chardin's work. *The Phenomenon of Man* was not published in English until in 1959, and I did not read it until in 1965, when the first draft of this essay was already completed, so any agreements or disagreements between us are purely coincidental. But, looking backward, I find that such areas are all the more interesting, and perhaps they may be so to the reader. Here I must confess that I was brought up, at the University of Chicago, on the "hard-nosed" version of the theory of evolution—the strictly objective, supposedly nonevaluative reconstruction of the past "facts," absolutely devoid of such teleological notions as "progress" or "goal" or "higher" or "lower" forms of life, and strictly confined to "earlier" and "later" and "more simple" and "more complex" and "random changes" and "natural selection" and "survival of the fittest" as descriptive-explanatory devices. But even in those days some students betrayed their secret teleological cast of mind by asking why, if mere survival is the goal, should we go beyond the amoeba, which is immortal? The answer was, random mutations, caused by cosmic rays, gamma rays from radioactive materials, and other instances of "blind chance." Change cannot be stopped, and we should be more surprised at relatively stable forms than at change. This made me an early Heraclitean.

My first reactions to *The Phenomenon of Man* were mostly negative. I kept saying to myself, "Surely this paleontologist must know *how* evolution operates!" But I could not deny a certain admiration for the way he invented new categories and then *reclassified* great masses of accepted "facts" under these new categories. As Camus said, reading Teilhard is like looking through the wrong end of a telescope. Not only is a thousand years as a day, millions of years are covered in one word. And after I read *The Appearance of Man* and *The Future of Man* and some of Teilhard's religious writings, it began to dawn on me that Teilhard was engaged in what I have called in Chapter Seven "totality mapping." As I noted there, the greatest danger in totality mapping comes from the vacuousness and vagueness that appear in the concepts themselves when they are taken from regional mappings and stretched and extrapolated to "cover" totality. We do not wish so much to agree or disagree with them as to ask, *what* is really being said? Furthermore, all the regional mappers are likely to boggle at this stretching of their concepts, for it was precisely by limiting their fields that they were able to make their concepts adequate to the subject matter.

The first difficulty, then, is the level of abstraction on which Teilhard operated, which permitted or forced him to use the word "evolution" to cover a sweep of time covering not only the sidereal universe in its development but the biological prelude, plus the prehistorical, the historical, the present, the near-future, and the far-future "story of mankind." He also used the better word "cosmogenesis" to cover this long stretch of time, and he invented the word "Christogenesis" to take care of his belief that Christ is the goal of cosmogenesis. His overall argument is really very simple: If there is a process that has produced the physical cosmos, if this process finally managed to produce life, if the biological process finally managed to produce man, and if the cultural process finally managed to produce Christ—assessed as the goal for the whole works—then how can we object to saying that the beginning and end must be part of the same totality process and, hence, cosmogenesis must be at the same time Christogenesis?

We may not object, but what do we gain? Surely *at the same time* cannot possibly mean the *same* process. Biologists would object to the implication that the process that produced the atoms and the stars is the *same* one that produced the different species on this planet. Historians would by and large object to the idea that the process that produced the history of civilizations is identifiable with "the survival of the fittest," although there some Social Darwinists might be willing to stretch a few points. Even Teilhard did not think that the "survival of the fittest" would produce any Christians. But once we have *Nous*, the human form of reason, a different process—the cultural process—is initiated, and it is different from but not discontinuous with, the previous process. Perhaps that is what the word "same" is intended to convey in Teilhard's view—that the total universe process consists of a number of parts in a series, each part different from but not discontinuous with, the antecedent and subsequent part. The total effect is cumulative, the later parts containing the earlier parts in overlapping and somewhat modified form. And man is, at this point in the total process, the summit and summation of the overlapping, cumulative processes that brought him here—and, the reasoning continues, he is by that very fact seen to be unfinished and ready for the next part of the process.

Teilhard described a number of these overlapping stages leading from the animal to man-the-brainy-ape to man, the future consummator of evolution, calling them by names so quasi-physical and quasi-biological that one is not sure, at least I am not sure, what they mean. These names are usually constituted by combining some common scientific or everyday term with the Latinizing endings "-ization," "-ification," or "-ation," which instantly converts a thing or action or feeling into a process; usually it is a slow, long-range process that could be "observed" only if in imagination the observer removed himself a great distance from earth and could survey enormous lengths of time "at a glance." For example, cerebralization, hominization, planetization, unanimization, amorization, Christification, divinization. Here I must confine myself to the first two.

The term "cerebralization" represents Teilhard's argument against those biologists who refuse to concede that one successful form of life is better than another, since all that are here today, having survived, are by definition "fit." And "who is to say that a mammal, even a man, is more advanced, more perfect, than a bee or a rose?" But if we use the elaboration of nervous systems as the criterion of advance, "Not only does the arrangement of animal forms according to their degree of cerebralization correspond exactly to the classification of systematic biology, but it also confers on the tree of life a sharpness of feature, an impetus, which is incontestably the hallmark of truth. . . ." (1) To Teilhard, this differentiation of nervous tissue "*provides a direction*; and by its consequences, *it proves that evolution has a direction*." (2) Now the theory of knowledge proposed in my own essay would certainly lend support to Teilhard's feeling that cerebralization is the *direction* of animal evolution, but only because the theory spells out the *survival value* of the nervous system's "being a classifying machine in a classifiable world," and using the animal value system as an action-guidance system for seeking the beneficial and avoiding the harmful situation. For a given environmental niche and a given gene pool, the better brains would tend to out-survive the poorer brains, thereby contributing the genes of their better brains to the genetic composition of that particular pool. But I am not even sure that Teilhard would have welcomed this support, since he appears to be talking about something other than survival value, some thing more like "matter reaching out toward consciousness." He said that in order to see a direction in evolution as an increase in consciousness-complexity that accompanies cerebralization, we have to move from the *without* to the *within* of things. I agree, and the primary emotion repertoire in species-specific behavior patterns describes the *within* of animals. But then Teilhard was willing to extrapolate this *within* clear down through the plants and even the molecules and atoms, thereby creating what sounds to me suspiciously like animistic botany and animistic physics. I draw the line at the plants. Granted that a certain amount of consciousness (which I called protoeuphoria and protodysphoria) accompanies every degree of elaboration of the nervous system—by the same token, since plants have no nervous system they have no consciousness, no *within* of things.

I am not even certain what the term "hominization" includes. Let us assume (using Teilhard's diagram on p. 121 of *The Appearance of Man*) that it includes at least the period from prehominian types (−500,000 years) to Neolithic *Homo sapiens* (−20,000 years). Teilhard thought that being human must include reflective, self-conscious thinking, and surely Cro-Magnon the cave decorator must have been capable of that. If we could only "read" the mythologies on the walls of Lascaux and Altamira, what would we not learn about what he feared and what he admired, what he thought of himself and of the "powers" in the world around him? For instance, why are the men in these paintings always so small and stick-like, and the animals so large and glorified? Did the cave dwellers worship the animals they were forced to hunt and destroy? Did they perhaps ask their pardon for having to kill them? Or did they perhaps try to propitiate the souls of the animals in order to insure a successful hunt? Did they try to hide from the revenge of the animals by representing the men as small and of no account? And when we come to burial customs, we certainly encounter consciousness of death and reflection on its meaning. Like Teilhard, I have nothing but the profoundest sympathy and admiration for Neolithic man in his late Upper Pleistocene setting.

But we have to talk about man in his twentieth-century setting, in particular the biblical view of modern man, and I am afraid that the theory of human knowledge here presented would not give much support to Teilhard's overly optimistic speculations. For a Christian, he seemed to be curiously unaware of original sin—how it enters into the very structure of knowledge and into the supposedly human value systems that direct how the knowledge is to be achieved and used, and for whose benefit. What is original sin anthropologically, if not the prolongation and intensification of Neolithic behavior patterns far, far beyond the point where they are necessary to survival? In the long process of hominization, where does responsibility begin? Our own creation myth tells us that it does not begin until *after* man achieves conscious knowledge of good and evil. Adam before the fall did good and evil, but he was exonerated by his innocent ignorance of what he was doing to others, just as animals and children are so exonerated today. The animal value system is surely inhuman, but let us not forget that no animal, not even the most ferocious lion, is capable of anything like cruelty or malice of forethought, because of his limited imagination and foresight. *After* the fall, Adam *knows* what evil he is doing to others. Thus from the purely animal point of view he would have to be regarded as the most foresightfully vicious, intelligently cruel, and egotistically exploitative animal in the entire animal kingdom. But there are mitigating circumstances, for Adam never would have survived if he had not also become the most foresightfully protective, intelligently cooperative, and individually self-sacrificing animal with regard to his own family and tribe.

Next after hominization come planetization, unanimization, and amorization, and if I understand him correctly, Teilhard conceived the last two after the model of the unanimity of scientific goals and the fellowship felt by scientists engaged in a common good cause—a communion of saintly scientists. Surely that is an attractive ideal, but when is it likely to happen? (Even today, the far-from-saintly scientists form a tribally organized power elite, while the rest of humanity is being born faster than it can be made literate.) Would it not be far more realistic, and therefore hopeful, to start with something that even today is absolutely universal in the human family and is reborn in every generation—the primary emotion repertoire as patterned for survival in Neolithic times—and try to modify *that*, culturally, with the aim of leading toward the viable global village, yes, the viable global *tribe*, unified in trying to save the planet? Of course it wouldn't be the kingdom of God, but it might buy up the time, the precious time, needed for the ultimate redemption of mankind.

In the long-range, overall view, I would *like* to agree with Teilhard. The close-up vistas are misty, however, and as I peer through the fog, I think we must be somewhere between planetization and amorization—and with our hominization barely begun! Perhaps if we were to correct our theology a bit, we could see more clearly what our immediate tasks ahead must be.

REFERENCES

1. Pierre Teilhard de Chardin, *The Phenomenon of Man*, Wm. Collins Sons and Co., Ltd., London, 1959, pp. 145–146.
2. Teilhard de Chardin, *The Phenomenon of Man*, p. 146.

Appendix Two The Author's Basic Convictions and Major Cultural Influences

I

I graduated from the University of Chicago in 1937 with a B.S. in physics. Prior to majoring in physics, however, I had been much absorbed in neurophysiology, where my great sources of inspiration had been Ralph Waldo Gerard, one of the pioneers in "brain waves," and the physiologist Anton J. Carlson. Thanks to the New Plan then in effect at that university, I had also become a "well-rounded person." I had always intended to go back to brain physiology, after mastering the necessary techniques for electrical measurements, circuitry, and so on, but alas, life and history intervened. I married my physics teacher, moved to Rutgers University, and then came World War II.

I had been brought up by my parents, who were both Czechs emigrated from Vienna in 1912, in the standard anticlerical atheism of most of the liberal intellectuals of Europe at that time; in my college days nothing could have been farther from my thoughts than religion. Like most scientific humanists

I had assumed that the answers to the fundamental problems of our existence were obvious: more science for our technical problems, and more education for our moral problems. The rise of the new barbarism in the very country most admired for its scientific and educational advances caused me to initiate a radical overhauling of my philosophical underpinnings, which had been logical positivist in relation to science but now had become mostly confused in relation to everything else.

I was looking for a concept of man broad enough to include not only ideas of the nature of science but also ideas of what man ought to do with his science. It was in the cool, undogmatic, friendly-to-science atmosphere of the Quakers that I first could begin to tolerate the Bible and other religious writings as a possible source of human wisdom. In 1945–1946 I attended Union Theological Seminary to study under Paul Tillich and John T. McNeil. I read extensively in Barth, Brunner, Kierkegaard, and European existentialism. I also rediscovered literature and the other arts, although both my parents had been "artistic"—my father a sculptor, my mother an operatic singer. I myself had practiced painting and music since childhood, but I had never thought about them other than as the "trimmings" of the cultured life. My early training in physiology and physics would not lie quiet, however, and what really bothered me during the mid-1940s was that so many of the existentialists, who could be so profound about human life, kept saying that "science is meaningless," whereas the scientists, shut up in their narrow specialties, kept accusing the existentialists of "irrationalism" and "emotional tantrums."

In 1948 we moved to Kenyon College, and it was through knowing the poet John Crowe Ransom, reading the *Kenyon Review*, and digesting the New Criticism with the assistance of the Rev. Charles Joseph Stoneburner and others, that I first saw a glimmer of light on the problem of escaping the positivistic trap, the dichotomy: cognitive/emotive. It would require a cognitive theory of poetry and art, nothing less. The writings of Susanne K. Langer and R. G. Collingwood were of great assistance to me in formulating my own ideas of the kind of knowledge that is available through the arts; but between them they did not have either a sufficiently universal theory of the imagination or a connecting link between the "fine arts" and the ordinary, everyday emotions that permeate the actions and attitudes of even the most unartistic people. Both Langer and Collingwood were against psychology, which was still dominated by Freudian interpretations of art and literature.

Kierkegaard had a great influence on me, so much so that I wrote a book on him, called *In Search of the Self: The Individual in the Thought of Kierkegaard*, published in 1962 by Muhlenberg Press, now Fortress Press. The *self* as a process of becoming, going through the "stages on life's way"—first the hedonic, then the ethical, the religious, and last the Christian—impressed me with the difficulty of Christianity when it is "inwardly appropriated" and not just a matter of mouthing formulas. The difference between words and deeds (i.e., that words are cheap and actions are expensive); the delusiveness of thinking about man in abstract, systematic terms, which seemingly avoid decisions; and the importance of decisions, which determine not only what I do but the self I thereby become—all these I learned from Kierkegaard. If every self is a hierarchy of value priorities, then his top priority is every individual's conscious God or his unconscious God-surrogate, which he betrays in his actions or inactions, whatever his thoughts or words. The inescapability of religion thus expressed is a whole psychology of religion which also explains the religion of so-called secularists and atheists of our times. And finally, there is the "absurdity" of the "existential situation"—namely, that we are forced by life to act in situations so complex that we can neither predict what would be the outcome of actions *A, B,* or *C* nor decide which outcome we ought to desire, in view of *its* further consequences; thus we are always thrown back on faith and always in need of repentance, forgiveness, and another chance. This was Kierkegaard's protest against the idealistic rationalists, who advised man, in Kant's terms, to act always so that the principle of his action could be erected into a universal law. To a biologist, the Hegelian view of the world as the march of Absolute Reason through history was so ridiculous that I wholeheartedly admired Kierkegaard's struggle against it, as well as the striving of other existentialists and phenomenologists for a more realistic appraisal of man's actual situation in the world.

Concurrently, I had been reading the Bible with the assistance of many interpreters and exegetes, and I was increasingly attracted to its realistic view of man, especially in the Old Testament. Of course it was also full of primitive ideas, such as the inheritance of acquired characteristics (the tribe of so-and-so shall earn their living by such-and-such) and patriarchal tribal arrangements of society. But what ancient writing was not? Is not the Bhagavad-Gita based on the Hindu caste system, and does not Lord Krishna finally advise Prince Arjuna to cast his doubts on Brahma and perform the duties of his station? A cognitive theory of art throws an enormous light on the mythologies of both "primitive peoples" and advanced religions, including our own, in disclosing man's behavior as an inescapably religious animal. The Hebrews attacked the mythology of the neighboring tribes, but only to proclaim the superiority of their own, and the sovereignty of Jahweh, who is "above all gods," above all "nations," above all tribes. Well then, why did the ancient Hebrews fail to evangelize their neighbors? They were too tribalistic themselves, and, although there are appeals for universalism in some of the prophets and psalms, it remained for Christianity to put universalism into evangelistic practice by proclaiming Christ the redeemer of *all* mankind. We cannot here go into the historical falling-short of *that* ideal.

It was over the question of redemption that I came to a parting-of-the-ways with Buddhism. The Quakers had taught me to be tolerant of all religions and especially sympathetic to the mystical tradition in most of them. I was also a great admirer of Gandhi's "Satyagraha"—his political technique of passive resistance, or rather, nonviolent noncooperation. And I could even appreciate that withdrawal from the senses and the desires as a preparation for Nirvana would constitute a self-liberation from slavery to the senses and the desires that would come in handy for the achieving of any goal, provided one could still have a goal in the state of apathy that would follow. I found Nirvana altogether too easy to arrive at, but always at the expense of "losing the world." How is it possible to get into a state of complete detachment from the world and still have compassion for one's fellow man, to desire *his* good, for instance, his feeding, clothing, and freeing from oppression? The Buddha himself had been moved by compassion, to free men from suffering. Also, my biological training would not permit me to take so negative a view of the senses and the emotions, since in some forms they are necessary for survival. Furthermore, if the ultimate goal of life is to secure release from the cycle of rebirths and redeaths, would it not be better never to have been born at all? I was not prepared for such a negative metaphysical verdict on life itself, nor for the view that the joys and sufferings of this life were "illusions," "Maya," mere "nature," as against Pure Being. Some kinds of suffering are an early warning system against danger, other kinds are a means to achieve a goal, still others are goads to cultural creation. The problem for redemption on this planet is how to overcome, or how to deal with, useless, unnecessary suffering. However, I still believe there is a great deal we in the West can learn from Buddhism about being less aggressive and competitive in the pursuit of our goals.

I freely confess that I may have missed a great deal in the more modern philosophical developments of Buddhism, especially the Mahayana type, but I became increasingly interested in the biblical view of man, which was both realistic about man's sinfulness and hopeful about his ultimate redemption.

Christians pray, Thy will be done *on earth* as it is in heaven. Heaven I could understand as the repository of those products of the imagination we call not-yet-realized but possible and desirable ideals or goals for human kind (greater truth, love, justice, freedom from oppression, etc.). Heaven as the repository for human souls after death leads to the same impasse as Nirvana—it leaves the problems of historical earthly life untouched; and as a reward for living a good life, it makes even true virtue impossible by entangling it in the reward-disease (see Kierkegaard's *Purity of Heart*). The notion of Heaven as the Kingdom of God on earth at some future time but even now partially present, makes the most sense of all, since every individual, in his attitudes, actions, and the consequences of his actions, must inevitably help or hinder its arrival or degree of realization. He can only choose, consciously, to try to help or hinder, in however small a way. (I cannot begin to name the heresies of which official Christian theology could accuse me, but then, it is much in need of house cleaning itself, or shall we say, museum cleaning?)

II

Here I can set down only those aspects of the biblical view of man, as I understand it, that influenced me in the writing of this book. First there is the inescapability of religion, as reflected in the biblical fight against idolatry. The choice is not between God or no God, or between one God and several gods, but between God above all gods and the lesser gods, who may be powerful and even worthwhile but who become idols when they usurp the place of the Most High. This kind of talk I regard as the proper mythological way of speaking about the totality of the universe in both its immanent and its transcendent aspects. So-called atheists are worshippers of science, art, power, money, their nation, some political system, or themselves. Secondly we have the covenant relationship between God and man. This pins both God and man to history. God promises to be *with* man, if man will learn what God is trying to teach him in history and by means of history. The covenant eliminates all ahistorical views of heaven and earth, including the Platonic heaven of eternal, unchanging, perfect ideas, of which this earth and men are poor imitations. Also eliminated are mathematical metaphors for necessity, which negate both divine and human freedom.

How does an evolutionist read the properly mythological language of Genesis? Well, one way is to say that a million years are like a day in the eyes of God and to proceed with the literal interpretation, as if nothing had changed. A much more profitable way is to see Genesis in the context of cultural anthropology, which welcomes all origin stories, not for how much they agree with modern views but for what they reveal of ancient man—his curiosity about his own past, present, and future; his "feeling of dependence" on some immensity much greater than himself; and his imaginative filling-in of that immensity with concrete "powers" that can be approached, propitiated, mollified. For why should an evolutionist, of all people, deny that human existence is both precarious and dangerous and that increasing consciousness simply increases the need for securing benefits and warding off threats, as well as satisfying curiosity?

To me, what distinguishes the Genesis story from other origin stories is that, instead of providing an assortment of gods and demons to account for good and evil in the world, it zeroes in on the origin and meaning of *responsibility* in that large-brained creature, *Homo sapiens*. The turning point is *knowledge of good and evil*—consciousness that we do both good and evil to others in order to survive, whereas the other animals do their good and evil in blissful ignorance, not conscious of the consequences to others. Of course the writers of this story knew nothing about evolution or the survival of the fittest as the means by which this species was created, but they had sufficient insight into the dubious quality of knowledge to present it first as a temptation, and then as a burden for which certain penalties must be paid. Perhaps they watched in their own children the same transition from thoughtless happiness to the grownup's worried conscience. They also seemed to have observed that there is something disproportionate about the evil that men do, as in the story of Cain and Abel. Here there is no question of survival, only sibling rivalry, or only Cain's hurt feelings at the father's misprizing of his offerings, and after the murder, Cain's refusal of responsibility. The story may have been put there as a warning for fathers or as a paradigm of the good father, since God punished Cain but did not utterly abandon him. (God put the "mark" on Cain to protect him from other people, as yet nonexistent. What do the literalists make of that?) In this compendium of origin stories, God also punished Adam and Eve—but for what? For knowing good and evil? For thinking to become immortal thereby? Or for misusing this knowledge? Mythologically, God must adjust himself to this *fait accompli* of his creature. He repents and starts all over again, but just before the second try, it is the imagination and the heart that are blamed for evil doing. "The Lord saw that the wickedness of man was great in the earth, and that every imagination of the thoughts of his heart was only evil continually." (Genesis 6:5)

Since for my cognitive theory of art I needed a fairly comprehensive theory of the imagination and the emotions, I was much interested in this discovery. In the Bible the word "imagination" is used mostly in the negative sense of "vain imagination." For instance, Paul used it to describe the pagan's lack of true knowledge of the divine creator:

> For the invisible things of Him from the creation of the world are clearly seen, being understood by the things that are made, even his eternal power and Godhead; so that they are without excuse; because that, when they knew God, they glorified him not as God,

neither were thankful: but became vain in their imaginations, and their foolish heart was darkened. (Romans 1:20–21)

"Imagination" is also used simply in the sense of "thoughts" or "speculations," but again in the negative sense of evil, secret planning in the heart. Now, the double meaning of the word "vain" as both futile and conceited is very handy if you are looking for something to label the knowledge-hindering functions of the imagination, but it is very difficult to find a single word for the opposite, knowledge-enhancing functions. Disciplined? Teachable? Realistic? What is the opposite of both futile and conceited? I tried many words, and finally gave up, in favor of describing specific functions.

Since I had already dismissed, both on biological and on philosophical grounds, any overly ascetic or world-escaping views of man, I was particularly impressed by the "wholeness" of the biblical view of man and by the cognitive implications of this wholeness. Certainly there are isolated instances of asceticism, but no systematic downgrading of either the senses or the feelings and desires, and this is reflected in biblical man's ideas about knowledge. In the Bible, the knowledge available to the creature man is no guarded storehouse of divine secrets, no arcane theosophy or compendium of magical religious practices; it is ordinary, everyday human knowledge such as comes to men from ordinary experience in making a living and getting along with one another. Such knowledge is sometimes called by us, somewhat pejoratively, "practical" as against "theoretical" knowledge. In the entry on Knowledge in Alan Richardson's *Theological Wordbook of the Bible* we read that in the Old Testament

The Hebrew *yada'* means know (whether a person or thing), perceive, learn, understand, have skill; and also, with a wider sweep of meaning than our "know", experience good or bad. . . . In the case of God, knowledge means his providence and prosecution of his good purposes; and particularly his choice of a man or nation to play a part in those purposes. . . . For the Rabbis of later Judaism, knowledge was primarily knowledge of the Law. (1)

Naturally the problem known as philosophical scepticism does not even arise in the Bible. The questioning of the testimonies of one's own senses, because of their privacy or their changeability, never occurs. Unlike us, the men of the Bible not only trusted their senses, they did not separate their senses from other activities of the whole human being (as we do) from thinking, willing, feeling, evaluating. They could speak of good as "sweet" and of evil as "bitter." Mind, heart, soul, spirit, are not functionally abstracted and separated as with us. "In particular, the widely held distinction between mind as seat of thinking and heart as seat of feeling (esp. tender feeling) is alien from the meaning these terms carry in the Bible." (2) This wholeness of man, whether mind, soul, spirit, or heart, refer to the person's *aliveness*, as against his deadness, or inactivity.

> "The Lord stirred up the spirit of Cyrus," i.e., moved Cyrus to do something. . . . "Call to heart" is one way of saying "remember." "To set one's heart on" means to give attention and the phrase "with the whole heart" (often plus "with the whole soul") means full and undivided attention and devotion. (3)

Because of their yet unfragmented wholeness, the individuals in the Bible collide and interact with nature, with one another, and with God, in a vivid, unjaded atmosphere that strikes us, at this remove, as resembling the naive realism of the child.

In the Bible there appears quite a bit of practical, common-sense knowledge of what we would call nature, the kind of knowledge required even today by herdsmen, farmers, builders, and craftsmen. But this knowledge is always seen against a background of the mysterious power and wonder in nature, which fill us with fear and awe because they represent

God's having a hand in it, His being the master over His own creation. Knowledge and mystery always go together, as do knowledge and piety. There are things that can be learned, the results of experience, handed down from father to son, or from master to apprentice. But there are other things that can only be wondered at. Typical is the attitude toward the agricultural cycle in the Bible. It is a learnable fact that the ground must be prepared and the seed must be put in the ground at the most favorable season. But then follows a period, while the seed is in the ground and nothing at all seems to be happening as far as the outward eye can see, which is always regarded with great wonder and pious gratitude, since it must be regarded as God's secret workings directly upon the seed.

Of course there was always the temptation, increasingly pressing as the nomadic tribes settled into the agricultural pattern of Canaan, to supplement and bolster up this very limited knowledge of nature by adding to it the fertility-cult practices of the neighboring Canaanites. "People believed that the agricultural harvest would not be plentiful unless the fertility powers were worshipped according to the ways of Canaan. To have ignored the Baal rites in those days would have seemed as impractical as for a modern farmer to ignore science in the cultivation of the land." (4) The prophets, however, regarded such a play-it-safe attitude as an insult to Jahweh. As if the Lord of creation, the redeemer of Israel and the judge of all nations, could not also take care of the processes of nature!

> *Behold, I will make thee a new sharp instrument having teeth: thou shalt thresh the mountains, and beat them small, and shalt make the hills as chaff.*
>
> *Thou shalt fan them and the wind shall carry them away, and the whirlwind shall scatter them: and thou shalt rejoice in the Lord, and shalt glory in the Holy One of Israel.*
>
> *When the poor and needy seek water, and there is none, and their tongue faileth for thirst, I the Lord will hear them, I the God of Israel will not forsake them.*

> *I will open rivers in high places, and fountains in the midst of the valleys: I will make the wilderness a pool of water, and the dry land springs of water.*
>
> *I will plant in the wilderness the cedar, the shittah tree, and the myrtle, and the oil tree; I will set in the desert the fir tree and the pine, and the box tree together:*
>
> *That they may see, and know, and consider, and understand together, that the hand of the Lord hath done this, and the Holy One of Israel hath created it.*
>
> Isaiah 41:15–20

In the Bible it is also assumed that men can have knowledge of other men, of what is in their hearts and minds. The technical problem known in philosophy as solipsism—that a man can know only the contents of his own mind and never the contents of another man's mind (and therefore cannot even be sure that other men exist)—is a problem that conspicuously fails to agitate the people in the Bible. To be sure, a man does not know perfectly what is in another man's mind and heart, and he may be deceived about a particular item; but this only means that he needs more knowledge to correct his ignorance, not that this kind of knowledge is in principle impossible. The Bible is also aware that man is always to some extent a mystery to himself, so why should he not always be to some extent a mystery to others? Again we have the mixture of knowledge and mystery, as in the case of nature. But in the case of knowledge of man, an entirely new element enters the picture, the possibility of willful ignorance, or, as the Bible calls it, blindness, stiffneckedness, and hardness of heart.

In the Bible, knowledge of other men and the responsibility it brings are directly related to each other and to the knowledge of God. As far as the knowledge of God is concerned, the Bible supports what we now call a revelation theory. That is to say, man can know God only because God has revealed himself to man, and only to the extent that God has revealed himself to man. But the Bible everywhere regards the results of God's self-revelation to man as *knowledge* of God, not as hysterical visions or private fantasies. Once the revelation has been made, knowledge of God of a certain degree is available to man, who thereafter cannot escape from its consequences by saying that knowledge of God is impossible. But again, as in the case of knowledge of man, there enter into the picture the elements of willful blindness, stiffneckedness, and hardness of heart. As the prophets repeatedly maintained, there is actually much more knowledge of God available to man than he is willing or eager to learn, for this kind of knowledge strikes at the heart in order to bring about a change in the manner of life, to lay *burdens* on the conscience. The knowledge of God teaches not only the history of the Jews as the mighty acts of God, but it covers also the field which we now call the realm of values. Whenever it is asked, what is good? What is bad? we need only answer by reciting the tablets of the law, or some convenient summary: "He hath showed thee, O man, what is good; and what doth the Lord require of thee, but to do justly, and to love mercy, and to walk humbly with thy God?" (Micah 6:8) The "values" thus proclaimed constitute knowledge of what God wants and does not want for man, not merely some private person's private inclinations.

This brings us to the phenomenon of conscience. In the Bible man is believed to be outwardly accountable to his fellow men in those local rules of society which particularize the idea of ethical obligation. In addition, he is both outwardly and inwardly accountable to God, who "sees" into his heart, for his feelings, attitudes, intentions, and motivations toward God and toward his fellow men. Both accountability and ethical knowledge are at the basis of the phenomenon of conscience, and you cannot have one without the other. Conscience is more than just the fear aroused by breaking a taboo, fear of the punishment that breaking the accepted mores of the tribe provokes, because conscience goes far beyond the areas of action specifically covered by such mores. Conscience, as this word is used in the Judeo-Christian tradition, presupposes the possibility of an agreement or disagreement between two kinds of knowledge—knowledge of what is, and knowledge of what ought to be. The word conscience means "co-knowledge," knowing together; but it would perhaps be better put as "knowing at the same time." A bad conscience means knowing at the same time that, in a given case, what should and could have been done contradicts or clashes with what actually was done; a good conscience means knowing that what ought to have been done actually was done. Accountability is the obligation laid on man, proximately by other men, but ultimately by God, to bring about an agreement between these two kinds of knowledge—between what is done and what ought to be done—but of course if it is maintained that ethical knowledge of any kind is impossible to attain, both conscience and accountability must vanish.

Nevertheless, in spite of this insistence that man can have some knowledge of God and of what God wants for man, and that he can even have a great deal more than he wants to have, nowhere is it even remotely suggested in the Bible that this knowledge of God is a Godlike knowledge of God, that it is omniscience, or perfection, in the cognitive sense. The Bible contains two opposite kinds of images of the knowledge of God by man; images of intimacy and images of remoteness. God is shown talking to man as another man would talk to him, as a father, as a king, as a judge, even as a husband to his bride. On the other hand, God is shown withdrawn into his mystery, inscrutable, untouchable in his holiness, hiding himself from the demanding, self-seeking eyes of men, wearied by their unworthiness. He is both immanent and transcendent, the still small voice, and the *mysterium tremendum et fascinans*. Only among idealistic philosophers is potential omniscience evaluated as a desirable goal for man, and even as a state of bliss. No man shall see God and live. This means not only that no man can claim omniscience, but that this power would destroy man if he could have it. To obtain an idea of the intolerable burden of omniscience, we must remember that with every increase of knowledge there comes a corresponding increase in responsibility and, therefore, eventually also an increase in guilt. This snowballing process could quickly drive a man mad, if he were not, as it were, protected and partially exonerated by the limitations of his cognitive equipment. (If anyone has any doubts on this process, let him remember a case like that of the discovery of the polio vaccine. Before this vaccine was available, being stricken by polio was like being struck by lightning—a part of the general precariousness of human existence, but scarcely an event for which guilt or blame could be assigned to any human agent. After the vaccine became available, being stricken by polio suddenly became a matter of someone's guilty irresponsibility, a sin of omission.)

I hope that this basically positive interpretation of the meaning of knowledge in the biblical view of man, which influenced me in attempting a unitary view of knowledge for modern man, will not be taken as an endorsement of a return to literalistic biblical fundamentalism, and still less as an endorsement of all kinds of Bronze Age hangovers of tribal, economic, political, and social structures, in which the Bible also abounds.

REFERENCES

1. Alan Richardson, Ed., *A Theological Wordbook of the Bible*, The Macmillan Company, New York, 1950, pp. 121–122 passim. Also with permission of SCM Press, Ltd., London.
2. Richardson, *Wordbook*, p.144.
3. Richardson, *Wordbook*, p. 145.
4. Bernhard W. Anderson, *Understanding the Old Testament*, Prentice-Hall, Inc., Englewood Cliffs, N.J., 1957, p. 100.

BIBLIOGRAPHY

Anderson, Bernard W., *Understanding the Old Testament*, Prentice-Hall, Inc., Englewood Cliffs, N.J., 1957.

Aristotle, *Poetics and Rhetoric*, Everyman's Library No. 901, J. M. Dent and Sons Ltd., London, 1953.

Aristotle, *Metaphysics*, Everyman's Library No. 1000, J. M. Dent and Sons Ltd., London, 1956.

Armstrong, A. H., *Plotinus*, Collier Books, New York, 1962.

Arnold, Magda, *Emotions and Personality*, Vol. II, Columbia University Press, New York, 1960.

Austin, J. L., *Sense and Sensibilia*, Oxford at The Clarendon Press, London, 1962.

Ayer, A. J., *The Problem of Knowledge,* Macmillan and Co. Ltd., London, 1956.

Barrett, William, *Irrational Man, A Study in Existential Philosophy*, Doubleday & Company, Garden City, N.Y., 1958.

Bartsch, Hans Werner, *Kerygma and Myth, A Theological Debate*, London, Society for Promoting Christian Knowledge, London, 1953.

Benda, Clemens E., *The Image of Love; Modern Trends in Psychiatric Thinking*, The Free Press of Glencoe, Inc., New York, 1961.

Berdyaev, Nicolas, *Solitude and Society*, Geoffrey Bles Ltd., London, 1947.

Black, Max, Ed., *Philosophical Analysis*, Cornell University Press, Ithaca, N.Y., 1950.

Blackmur, R. P., *Language as Gesture*, Harcourt Brace Jovanovich, Inc., New York, 1952.

Boman, Thorleif, *Hebrew Thought Compared with Greek*, The Westminster Press, Philadelphia, 1960.

Brain Mechanisms and Consciousness, A Symposium, Charles C. Thomas, Springfield, Ill., 1954.

Brock, Werner, *Existence and Being by Martin Heidegger*, The Henry Regnery Company, Chicago, 1949.

Brooks, Cleanth, *The Well Wrought Urn: Studies in the Structure of Poetry,* Harcourt Brace Jovanovich, Inc., New York, 1947.

Brown, Roger, *Words and Things*, The Free Press, New York, 1958.

Bruner, Jerome S., Jacqueline J. Goodnow, and George A. Austin, *A Study of Thinking*, John Wiley & Sons, New York, 1956.

Buber, Martin, *Eclipse of God; Studies in the Relation between Religion and Philosophy*, Harper & Row, New York, 1952.

Buber, Martin, *Between Man and Man*, Beacon Press, Boston, 1955.

Buber, Martin, *I and Thou*, Charles Scribner's Sons, New York, 1958.

Buber, Martin, *On Judaism*, Schocken Books, New York, 1967.

Bultmann, Rudolph, *Primitive Christianity in its Contemporary Setting*, Meridian Books, New York, 1956.

Bultmann, Rudolph, *The Presence of Eternity: History and Eschatology*, Harper & Row, New York, 1957.

Burke, Kenneth, *The Rhetoric of Religion*, Beacon Press, Boston, 1961.

Burtt, E. A., *The Teachings of the Compassionate Buddha*, The New American Library (Mentor Book), New York, 1955.

Chase, Richard, *Quest for Myth*, Louisiana State University Press, Baton Rouge, 1949.

Child Psychotherapy, Mary R. Haworth, Ed., Basic Books, Inc., New York, 1964.

Chisholm, Roderick M., *Perceiving: A Philosophical Study*, Cornell University Press, Ithaca, N.Y., 1957.

Christian, William A., *Meaning and Truth in Religion*, Princeton University Press, Princeton, N.J., 1964.

Church, Joseph, *Language and the Discovery of Reality*, Random House, New York, 1961.

Collingwood, R. G., *An Essay on Metaphysics*, Oxford University Press, New York, 1940.

Collingwood, R. G., *The Principles of Art*, Oxford University Press, New York, 1958.

Cornford, Francis Macdonald, *Plato's Theory of Knowledge*, Routledge & Kegan Paul Ltd., London, 1960.

Crombie, I. M., *Plato's Doctrines, Volume II, Plato on Knowledge and Reality*, The Humanities Press, New York, 1963.

Dobzhansky, Theodosius, *The Biological Basis of Human Freedom*, Columbia University Press, New York, 1956.

Dobzhansky, Theodosius, *The Biology of Ultimate Concern*, The New American Library, New York, 1967.

Eliade, Mircea, *Cosmos and History. The Myth of the Eternal Return*, Harper & Row, New York, 1959.

Eliot, T. S., *On Poetry and Poets*, Farrar, Straus & Giroux, New York, 1957.

Emmet, Dorothy M., *The Nature of Metaphysical Thinking*, Macmillan and Company, New York, 1945.

Empson, William, *Seven Types of Ambiguity,* Harcourt Brace Jovanovich, Inc., New York, 1931.

Existence, A New Dimension in Psychiatry and Psychology, Rollo May, Ernest Angel, and Henri F. Ellenberger, Ed., Basic Books, Inc., New York, 1958.

Federn, Paul, *Ego Psychology and the Psychoses*, Basic Books, Inc., New York, 1952.

Freud, Sigmund, *Civilization and its Discontents*, Anglobooks, New York, 1952.

Freud, Sigmund, *Moses and Monotheism*, Vintage Books, New York, 1955.

Friedman, Maurice, *The Worlds of Existentialism*, Random House, New York, 1964.

Fromm-Reichmann, Frieda, Ed., *Progress in Psychotherapy*, Grune & Stratton, New York, 1956.

Frye, Northrop, Ed., *Romanticism Reconsidered*, Columbia University Press, New York, 1963.

Frye, Northrop, *Fables of Identity: Studies in Poetic Mythology*, Harcourt Brace Jovanovich, Inc., New York, 1963.

Gellner, Ernest, *Words and Things, A Critical Account of Linguistic Philosophy*, Victor Gollancz, Ltd., London, 1959.

Gogarten, Friedrich, *Demythologizing and History*, Charles Scribner's Sons, New York, 1955.

Hampshire, Stuart, *Thought and Action*, The Viking Press, New York, 1960.

Harper, Robert A., *Psychoanalysis and Psychotherapy, 36 Systems*, Prentice-Hall, Inc., Englewood Cliffs, N.J., 1959.

Harrison, Jane E., *Ancient Art and Ritual*, Holt, Rinehart & Winston, New York, 1913.

Hatch, Edwin, *The Influence of Greek Ideas on Christianity*, Harper & Row, New York, 1957.

Hayek, F. A., *The Sensory Order*, The University of Chicago Press, Chicago, 1952.

Hebb, D. O., *The Organization of Behavior*, John Wiley & Sons, New York, 1949.

Heidegger, Martin, *Being and Time*, Harper & Row, New York, 1962.

Henderson, Ian, *Myth in the New Testament*, The Henry Regnery Company, Chicago, 1952.

Hick, John, *Faith and Knowledge*, Cornell University Press, Ithaca, N.Y., 1957.

Hopper, Stanley Romaine, *Spiritual Problems in Contemporary Literature*, Harper & Row, New York, 1952.

Horney, Karen, *Neurosis and Human Growth*, W. W. Norton & Co., New York, 1950.

The Later Heidegger and Theology, Vol. I, *New Frontiers in Theology*, James M. Robinson and John B. Cobb, Jr., Eds., Harper & Row, New York, 1963.

Jaeger, Werner, *Aristotle: Fundamentals of the History of his Development*, Oxford University Press, New York, 1948.

Jaspers, Karl, *The Perennial Scope of Philosophy*, Philosophical Library, Inc., New York, 1949.

Jenkins, Iredell, *Art and the Human Enterprise*, Harvard University Press, Cambridge, Mass., 1958.

Jesperson, Otto, *The Philosophy of Grammar*, G. Allen & Unwin Ltd., London, 1925.

Jonas, Hans, *The Phenomenon of Life: Toward a Philosophical Biology*, Harper & Row, New York, 1966.

Jones, Ernest, *The Life and Work of Sigmund Freud,* Vols. I, II, and III, Basic Books, Inc., New York, 1957.

Kermode, Frank, *The Sense of an Ending; Studies in the Theory of Fiction*, Oxford University Press, New York, 1967.

Kierkegaard, Søren A., *The Concept of Dread*, Princeton University Press, Princeton, N.J., 1944.

Kierkegaard, Søren A., *The Sickness Unto Death*, Princeton University Press, Princeton, N.J., 1944.

Krieger, Murray, *The Tragic Vision*, The University of Chicago Press, Chicago, 1960.

Kubie, Lawrence S., *Neurotic Distortion of the Creative Process*, University of Kansas Press, Lawrence, Kansas, 1958.

Kubie, Lawrence S., *Practical and Theoretical Aspects of Psychoanalysis*, Frederick A. Praeger, New York, 1960.

Kuhn, Helmut, *Encounter with Nothingness: An Essay on Existentialism*, The Henry Regnery Company, Chicago, 1949.

Langer, Susanne K., *Philosophy in a New Key*, The New American Library, New York, 1948.

Langer, Susanne K., *Feeling and Form: A Theory of Art*, Charles Scribner's Sons, New York, 1953.

Langer, Susanne K., *Reflections on Art*, The Johns Hopkins Press, Baltimore, Md., 1958.

Lévi-Strauss, Claude, *The Savage Mind*, The University of Chicago Press, Chicago, 1966.

MacCallum, Reid, *Imitation and Design*, University of Toronto Press, Toronto, Ont., 1953.

Marcel, Gabriel, *The Mystery of Being*, Vols. I and II, The Henry Regnery Company, Chicago, 1950.

Matthiessen, Francis Otto, *The Achievement of T. S. Elliot: An Essay on the Nature of Poetry*, Oxford University Press, New York, 1959.

Michalson, Carl, Ed., *Christianity and the Existentialists*, Charles Scribner's Sons, New York, 1956.

Miller, Libuse Lukas, *The Christian and the World of Unbelief*, Abingdon Press, Nashville, Tenn., 1957.

Miller, Libuse Lukas, *In Search of the Self: The Individual in the Thought of Kierkegaard*, Fortress Press, Philadelphia, 1962.

Minear, Paul Sevier, *Eyes of Faith. A Study in the Biblical Point of View*, The Bethany Press, St. Louis, Mo., 1966.

Mowrer, O. Hobart, *The Crisis in Psychiatry and Religion*, D. Van Nostrand Company, Princeton, N.J., 1961.

Murray, Gilbert, *Five Stages of Greek Religion*, Doubleday & Company, Garden City, N.Y., 1951.

Otto, Rudolf, *The Idea of the Holy*, Oxford University Press, New York, 1958.

Paul, Leslie, *Persons and Perception*, Faber & Faber, London, 1961.

Piaget, Jean, *Play, Dreams and Imitation in Childhood*, W. W. Norton & Co., New York, 1962.

Pitcher, George, *The Philosophy of Wittgenstein*, Prentice-Hall, Inc., Englewood Cliffs, N.J., 1964.

Plutchik, Robert, *The Emotions: Facts, Theories, and a New Model*, Random House, New York, 1962.

Polanyi, Michael, *Personal Knowledge: Towards a Post-Critical Philosophy*, The University of Chicago Press, Chicago, 1958.

Raisz, Erwin, *General Cartography*, McGraw-Hill Book Company, Inc., New York, 1938.

Ransom, John Crowe, *The World's Body*, Charles Scribner's Sons, New York, 1938.

Ransom, John Crowe, *The New Criticism*, New Directions, Norfolk, Conn., 1941.

Ransom, John Crowe, *Poems and Essays*, Vintage Books, New York, 1955.

Read, Sir Herbert, *Icon and Idea: The Function of Art in the Development of Human Consciousness*, Harvard University Press, Cambridge, Mass., 1955.

Reid, Louis Arnaud, *A Study in Aesthetics*, The Macmillan Company, New York, 1954.

Rogers, Carl R., *Client-Centered Therapy*, Houghton Mifflin Company, Boston, 1951.

Ryle, Gilbert, *The Concept of Mind*, Barnes & Noble, New York, 1949.

Sartre, Jean-Paul, *Existentialism*, The Philosophical Library, New York, 1947.

Scott, Nathan A., Jr., *Rehearsals of Discomposure*, King's Crown Press, New York, 1952.

Sherrington, Sir Charles, *The Integrative Action of the Nervous System*, Yale University Press, New Haven, Conn., 1961.

Snell, Bruno, *The Discovery of the Mind*, Harvard University Press, Cambridge, Mass., 1953.

Stauffer, Donald A., *The Nature of Poetry*, W. W. Norton & Company, Inc., New York, 1946.

Strawson, P. F., *Individuals: An Essay in Descriptive Metaphysics*, Doubleday & Company, Garden City, N.Y., 1963.

Suttie, Ian D., *The Origins of Love and Hate*, The Julian Press, New York, 1935.

The Theology of Paul Tillich, Charles W. Kegley and Robert W. Bretall, Eds., The Macmillan Company, New York, 1952.

Tillich, Paul, *Biblical Religion and the Search for Ultimate Reality*, The University of Chicago Press, Chicago, 1955.

Toulmin, Stephen, *The Place of Reason in Ethics*, Cambridge University Press, London, 1950.

Toulmin, Stephen, *The Philosophy of Science*, Harper & Row, New York, 1960.

Van Der Leeuw, G., *Religion in Essence and Manifestation*, Vols. I and II, Harper & Row, New York, 1963.

Van Leeuwen, Arend T., *Christianity in World History*, Charles Scribner's Sons, New York, 1964.

Voegelin, Eric, *Order and History, Vol. I, Israel and Revelation*, Louisiana State University Press, Baton Rouge, La., 1956.

Wheelwright, Philip, *Metaphor and Reality*, Indiana University Press, Bloomington, Ind., 1952.

Whorf, Benjamin Lee, *Language, Thought and Reality*, Technology Press, Cambridge, Mass., 1956.

Wiener, Norbert, *The Human Use of Human Beings*, Doubleday Anchor Book, Garden City, N.Y., 1954.

Wilder, Amos N., *The Spiritual Aspects of the New Poetry*, Harper & Row, New York, 1940.

Wimsatt, W. K., Jr., *The Verbal Icon*, University of Kentucky Press, Lexington, Ky., 1954.

Wittgenstein, Ludwig, *Tractatus Logico-Philosophicus*, Routledge & Kegan Paul Ltd., London, 1922.

Wittgenstein, Ludwig, *Philosophical Investigations*, The Macmillan Company, New York, 1953.

Index

n refers to a footnote; q refers to a quotation